CHEMICAL ENGINEERING METHODS AND TECHNOLOGY

ANILINE

STRUCTURAL/PHYSICAL PROPERTIES, REACTIONS AND ENVIRONMENTAL EFFECTS

CHEMICAL ENGINEERING METHODS AND TECHNOLOGY

Additional books in this series can be found on Nova's website under the Series tab.

Additional e-books in this series can be found on Nova's website under the e-book tab.

CHEMICAL ENGINEERING METHODS AND TECHNOLOGY

ANILINE

STRUCTURAL/PHYSICAL PROPERTIES, REACTIONS AND ENVIRONMENTAL EFFECTS

KENNETH HERNANDEZ
AND
MARIA HOLLOWAY
EDITORS

nova publishers
New York

For permission to use material from this book please contact us:
Telephone 631-231-7269; Fax 631-231-8175
Web Site: http://www.novapublishers.com

Library of Congress Cataloging-in-Publication Data

ISBN: 978-1-62417-417-9

Published by Nova Science Publishers, Inc. † New York

CONTENTS

PREFACE

Being a precursor to many industrial chemicals, aniline's main use is in the manufacture of precursors to polyurethane. Like most volatile amines, it possesses the somewhat unpleasant odor of rotten fish. It ignites readily, burning with a smoky flame characteristic of aromatic compounds. In this book, the authors discuss the structural and physical properties, reactions and environmental effects of aniline. Topics include the green catalytic process for the synthesis of aromatic amines from hydrogenation of aromatic nitro compounds; aniline-based polybenzoxazine and their copolymers or composites; one-step C-H amination offers a new synthetic strategy towards the clean and effective production of aniline; synthesis and structures of gossypol mono-aldehyde derivatives with o-substituted anilines; and oxidative polymerization of diphenylamine-2-carbonic acid.

Chapter 1 - In the hydrogenation of aromatic nitro compounds, several poisonous intermediates such as nitrosobenzene, *N*-phenylhydroxylamine, azoxybenzene, azobenzene, and hydrazobenzene are commonly formed along with the desired product of aniline. The accumulation of these undesired intermediates should be avoided for producing high quality aromatic amines. Dehalogenation is another side reaction in the hydrogenation of halonitrobenzenes that produces hydrogen halide and thus reduces the atom economy of the reaction. It is also a difficult task to hydrogenate nitro group selectively in the presence of other reducible functional groups. Recently, many efforts are devoted to resolving these problems and some progresses have been achieved in developing efficient catalysts and new green reaction media. For example, the monometallic (Pt, Pd, Ru, Au, Ag, Ni, and Co) and bimetallic (Ni-Pt, Ni-Pd, Ni-Au, Pt-Au etc.) catalysts are extensively studied and discussed, the activity and product selectivity are improved significantly in some cases. In addition, green reaction media such as supercritical carbon dioxide ($scCO_2$), CO_2-H_2O, and CO_2-expanded organic liquids have also been studied for the hydrogenation of aromatic nitro compounds. The dehalogenation and the accumulation of harmful intermediates can be suppressed over Ni/TiO_2 in $scCO_2$, and the selective hydrogenation of water-insoluble nitrobenzene and chloronitrobenzene can be achieved over Ni catalysts in H_2O–compressed CO_2 system at 318-323 K. Moreover, the pressurization with CO_2 presents to vary the product selectivity in the nonpolar solvent but not in the polar liquid for the hydrogenation of nitrostyrene over Pt/TiO_2 in CO_2-expanded ethanol and toluene. Herein, the recent progresses in the hydrogenation of aromatic nitro compounds to aromatic amines will be reviewed. Particularly, some factors including the property of metal, type of support, and metal-support interaction will be discussed to reveal their effects on the activity and selectivity of catalysts. The effects of CO_2 will be discussed according to recent results of *in situ* Fourier transform infrared spectroscopy that disclose the interaction of compressed CO_2 with reactant, reaction intermediate, solvent, and the metal catalyst.

Chapter 2 - Polybenzoxazines as a novel phenolic type thermosetting polymer have been developed to overcome the shortcomings associated with the use of the traditional novolac and resol-type phenolic resins. They can be prepared from cyclic benzoxazines via thermally induced ring-opening polymerization. Various benzoxazine monomers can be synthesized from mono- or biphenols, paraformaldehyde and aniline. Then, the different approaches for performance improvement of polybenzoxazine have been investigated.

These approaches include hybrid with other high performance polymers and reinforcing with natural fillers, long fibers, and inorganic fillers. In these approaches, lowering of the cure temperature, enhancement of toughness, and improvement of thermal and mechanical properties were achieved. In addition, some application potentials of polybenzoxazine-based materials are also considered.

Chapter 3 - The C–N bond-forming reaction, by introducing nitrogen atoms into organic molecules, has an ongoing presence in research at the forefront of chemistry and interrelated fields. Among numerous investigations, the introduction of amino group (-NH_2) into inert aromatic C–H bond to produce aniline is utmost attractive and remains a significant challenge. In this chapter, the examples that concern the synthetic strategies towards aniline production from classical multiple-step method to modern one-step method are reviewed. The classical method for aniline production generally concerns nucleophilic substitution or reductive amination of pre-activated aromatics, but the modern one-step method deals with the direct transfer of -NH_2 group from an amino source, in particular ammonia (NH_3), to aromatic C–H bond. By shortening the synthetic steps, the direct C–H amination process becomes more simple, waste-minimal, and energy-conservative. In addition, insertion of -NH_2 group from simple inorganic aminating agent to aromatic C–H bond gives clue to the formation of organic amines that may be a crucial step to biosynthesis. The influences of several reaction parameters, including source of aminating agent, catalyst, oxidant, solvent, reaction temperature and pressure, are discussed. As a conclusion, the direct synthetic approach of one-step C–H amination is offering a new synthetic strategy towards the clean and effective production of aniline.

Chapter 4 - Gossypol mono-aldehyde derivatives with *o*-methylaniline, and *o*-hydroxyaniline have been synthesized and their structures have been studied by IR and ^{1}H NMR spectroscopic methods.

Chapter 5 - Polymers of diphenylamine-2-carbonic acid (DPhACA) were obtained for the first time by means of chemical oxidative polymerization in aqueous solution of ammonium hydroxide, a new polyacid, containing carboxyl groups which form intermolecular hydrogen bonds with aminogroups along the whole polymer chain, was obtained. The effect of polymerization conditions on the chemical structure and molecular-mass characteristics of polydiphenylamine-2-carbonic acid (PDPhACA) was studied.

According to IR spectroscopy data, while polymerization of DPhACA goes in ammonium hydroxide solution, growth of polymeric chain occurs due to C–C joining into 2- and 4- positions of aromatic rings relative to nitrogen. DPhACA polymers are amorphous and thermally stable.

In: Aniline
Editors: K. Hernandez, M. Holloway
ISBN: 978-1-62417-417-9

Chapter 1

GREEN CATALYTIC PROCESS FOR SYNTHESIS OF AROMATIC AMINES FROM HYDROGENATION OF AROMATIC NITRO COMPOUNDS

***Haiyang Cheng*[1], *Xiangchun Meng*[2,*], *Weiwei Lin*[1,3] and *Fengyu Zhao*[1,#]**

[1]State Key Laboratory of Electroanalytical Chemistry; Green Chemistry and Process Laboratory, Changchun Institute of Applied Chemistry, Chinese Academy of Sciences
[2]School of Chemical Engineering, Changchun University of Technology, China
[3]Graduate University of Chinese Academy of Sciences, China

ABSTRACT

In the hydrogenation of aromatic nitro compounds, several poisonous intermediates such as nitrosobenzene, *N*-phenylhydroxylamine, azoxybenzene, azobenzene, and hydrazobenzene are commonly formed along with the desired product of aniline. The accumulation of these undesired intermediates should be avoided for producing high quality

* Email: mengxiangchun@mail.ccut.edu.cn (X. Meng).
# Email: zhaofy@ciac.jl.cn (F. Zhao).

aromatic amines. Dehalogenation is another side reaction in the hydrogenation of halonitrobenzenes that produces hydrogen halide and thus reduces the atom economy of the reaction. It is also a difficult task to hydrogenate nitro group selectively in the presence of other reducible functional groups. Recently, many efforts are devoted to resolving these problems and some progresses have been achieved in developing efficient catalysts and new green reaction media. For example, the monometallic (Pt, Pd, Ru, Au, Ag, Ni, and Co) and bimetallic (Ni-Pt, Ni-Pd, Ni-Au, Pt-Au etc.) catalysts are extensively studied and discussed, the activity and product selectivity are improved significantly in some cases. In addition, green reaction media such as supercritical carbon dioxide ($scCO_2$), CO_2-H_2O, and CO_2-expanded organic liquids have also been studied for the hydrogenation of aromatic nitro compounds. The dehalogenation and the accumulation of harmful intermediates can be suppressed over Ni/TiO_2 in $scCO_2$, and the selective hydrogenation of water-insoluble nitrobenzene and chloronitrobenzene can be achieved over Ni catalysts in H_2O–compressed CO_2 system at 318-323 K. Moreover, the pressurization with CO_2 presents to vary the product selectivity in the nonpolar solvent but not in the polar liquid for the hydrogenation of nitrostyrene over Pt/TiO_2 in CO_2-expanded ethanol and toluene. Herein, the recent progresses in the hydrogenation of aromatic nitro compounds to aromatic amines will be reviewed. Particularly, some factors including the property of metal, type of support, and metal-support interaction will be discussed to reveal their effects on the activity and selectivity of catalysts. The effects of CO_2 will be discussed according to recent results of *in situ* Fourier transform infrared spectroscopy that disclose the interaction of compressed CO_2 with reactant, reaction intermediate, solvent, and the metal catalyst.

1. Introduction

The catalytic hydrogenation of aromatic nitro compounds is a common method to manufacture aromatic amines, which are important intermediates in the synthesis of dyes, pharmaceuticals, and agricultural chemicals [1]. The hydrogenation is usually carried out with heterogeneous metal catalysts, yielding aromatic amines through several possible pathways via some reaction intermediates such as nitrosobenzene (NSB), *N*-phenylhydroxylamine (PHA), azoxybenzene (AOB), azobenzene (AB), and hydrazobenzene (HAB) (Scheme 1) [2-4]. Generally, the rate determining step is the reduction of PHA to AN in organic solvent. NSB and PHA are reactive and tend to condense into AOB. Further hydrogenation of AOB leads to the AB and HAB compounds, which are finally transformed into aniline. The accumulation of these intermediates, especially PHA, is critical when the hydrogenation of NB is carried out in batch reactor at low or medium temperature [5]. Because these compounds are highly toxic and problematic, their accumulation during the reaction should be avoided for the safe and green production of AN [5, 6].

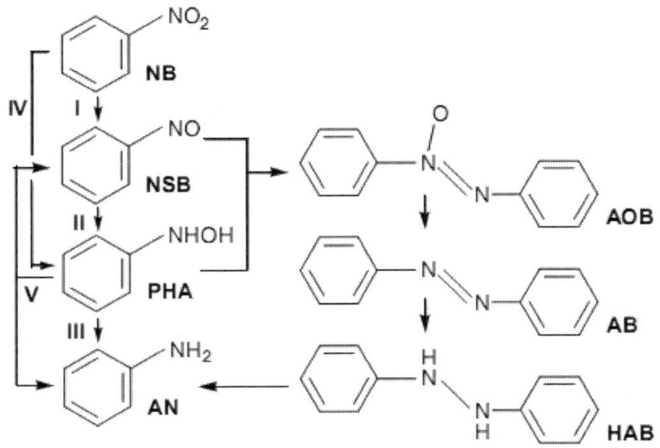

Scheme 1. Possible reaction pathways for the hydrogenation of nitrobenzene [2-4]. NB: nitrobenzene; NSB: nitrosobenzene; PHA: *N*-phenylhydroxylamine; AN: aniline; AOB: azoxybenzene; AB: azobenzene; HAB: hydrazobenzene.

Moreover, the production costs and waste minimization are of growing importance even for high value pharma- and agro-chemicals. Therefore, many efforts have been made to avoid the accumulation of PHA to achieve high yield of the desired amines by designing efficient catalysts and/or performing the reaction in green solvents, and some progresses have been achieved [7-11]. For example, the monometallic (Pt, Pd, Ru, Au, Ag, Ni, and Co) and bimetallic (Ni-Pt, Ni-Pd, Ni-Au, Pt-Au etc.) catalysts are extensively studied and discussed, the activity and product selectivity are improved significantly in some cases. Besides the accumulation of the intermediates, dehalogenation in the hydrogenation of halogenated aromatic nitro compounds usually happens, because the hydrogenolysis of carbon-halogen bond can be enhanced by amino substitution in the aromatic ring [12]. Many efforts have been made to prevent the dehalogenation by adding the additives [13] and modifiers [11, 14-16], for example, Wang et al. developed Pt/γ-Fe_2O_3 nanocomposite catalysts, with which the hydrodehalogenation of haloanilines was completely suppressed even at complete conversion of halonitrobenzene, and they proposed that the electron transfer from Pt nanoparticles to oxygen vacancies in the activated Pt/γ-Fe_2O_3 catalysts may play an important role in the complete suppression of the hydrodehalogenation of haloanilines [14, 15].

In addition, green reaction media such as ionic liquids, supercritical carbon dioxide ($scCO_2$), H_2O, and CO_2-expanded liquids (CXLs) were examined for the hydrogenation of aromatic nitro compounds and some achievements have been reported [7-9, 17-27]. For example, the dehalogenation and the accumulation of harmful intermediates can be suppressed over Ni/TiO_2 in $scCO_2$ [8], and the selective hydrogenation of water-insoluble nitrobenzene and chloronitrobenzene can be performed over Ni catalysts in a H_2O–compressed CO_2 system at 318-323 K [9]. The advantages of $scCO_2$ or the compressed CO_2 in hydrogenation of nitro compounds such as improving the reaction rate and adjusting the product selectivity have been well discussed according to the results from *in situ* Fourier transform infrared spectroscopy, which disclose the interaction of compressed CO_2 with reactant, reaction intermediate, solvent, and the metal catalyst [7, 8].

In this chapter, we give a review on the recent achievements in developing efficient catalysts and green catalytic systems for the hydrogenation of aromatic nitro compounds. Supported monometallic and bimetallic catalysts with good catalytic performance for the hydrogenation in organic solvents are reviewed firstly, and then the hydrogenation of aromatic nitro compounds under solvent-free conditions and in green media such as ionic liquids, $scCO_2$, CXLs, and water are described in detail. Finally, other important progresses in reduction of aromatic nitro compounds are described.

2. Heterogeneous Catalysts for Hydrogenation of Aromatic Nitro Compounds in Organic Solvents

Selective hydrogenation of aromatic nitro compounds has been investigated over different type catalysts such as supported metal (Pt, Pd, Ru, Au, Ag, Ir, Rh, Ni, Co) catalysts, polymer-protected metal nanoclusters, and organometallic complexes, etc. Supported noble metal (Pt and Pd) catalysts often exhibit high catalytic activity, and their catalytic performance is largely related to the metal specificity. For example, Pt-based catalysts exhibit high selectivity for the hydrogenation of halonitrobenzenes but moderate selectivity for the aromatic nitro compounds with reducible functional group (C=C, C≡C,

CN, C=O, etc.); Pd-based catalysts also exhibit high activity and selectivity without accumulation of the intermediates for the hydrogenation of aromatic nitro compounds; Ru-, Au- and Ag-based catalysts are selective for the hydrogenation of aromatic nitro compounds, but their catalytic activity needs to be improved; Ir- and Rh-based catalysts have received less attention because of their poor activity and selectivity. Non-noble metals like Ni and Co are abundant and cheap, and present moderate activity and high selectivity, so Ni- and Co-based catalysts have attracted more attention in the hydrogenation of aromatic nitro compounds.

The catalytic performance can be improved by changing the metal particle size, modulating the metal/support interaction, introducing some additives and alloying with other metals. The above related researches will be separately reviewed in the following several sections.

2.1. Pt Catalysts

The hydrogenation of aromatic nitro compounds over Pt based catalysts is generally faster than that over Ru, Au, Ag, Ir, Rh, and Ni based catalysts, but the accumulation of harmful intermediates usually occurs with Pt-based catalysts and many efforts for avoiding the accumulation have been made, such as alloying Pt with other metals (Ru, Pd, Au, Ni described in part 2.8) [28-33]. For the hydrogenation of aromatic nitro compounds with halogen (Cl, Br, I) substituent and other reducible functional groups (C=C, C≡C, CN, C=O etc.), the extent of dehalogenation with Pt catalysts is usually less than that with Pd catalysts. However, the selectivity needs to be further enhanced, even though some progresses have been obtained with Pt catalysts [13-16, 34-45].

Support may have great influence on the catalytic properties due to a strong metal-support interaction (SMSI) [62]. Some metal oxides such as TiO_2, CeO_2, SnO_2, Fe_2O_3, and Fe_3O_4 frequently show such SMSI effect when they are used as a support for the transition metals. For the TiO_2-supported metal catalysts, it has been clearly established that the suboxide TiO_x species ($x<2$) can migrate over the metal surface during the reduction at high temperature [16, 62]. Coq et al. reported that the TOF increased threefold in the hydrogenation of *p*-chloronitrobenzene (*p*-CNB) over Pt/TiO_2 reduced at high temperature (773 K) compared to that reduced at low temperature (523 K). They suggested that the TiO_x species on Pt polarize the N=O bond and make it highly susceptible to the hydrogen attack. Thus, the suboxide TiO_x

should be responsible for promoting the activity and selectivity in the hydrogenation of *p*-CNB over a Pt/TiO_2 catalyst in a SMSI state [34].

Table 1. Hydrogenation of aromatic nitro compounds with Pt-based catalysts

Catalyst	Substituent	Selectivity/ conversion/ yield (%)	TOF A/R [a] (s^{-1})	Ref.
0.48% Pt/TiO_2	Cl	--/--/99.3	--/7.1	[34]
0.3% Pt/TiO_2	Cl	--/99.7/90.7	--/1.45	[35]
0.2% Pt/TiO_2	C=C, CHO, CN, I, Cl, C≡C	>90/>95/--	--	[16]
$Pt/TiO_2/RGO$	H	99.9/100/--	16.4/--	[36]
$Pt/Fe_2O_3/C$	Cl	99.5/100/--	78/--	[37]
Pt/γ-Fe_2O_3	Cl, Br, I	99.9/>99.9/--	--	[14, 15, 38, 46]
Pt/Fe_3O_4	Cl, Br	>99.1/100/--	>2.0/--	[47]
Pt/Ni(C)	H	100/99/--	--	[48]
Pt/γ-ZrP	Cl	100/100/--	--	[39]
Pt/H^+-Atta	Cl	>99/100/--	--	[40]
Pt/MWNT	H	100/100/--	18.6/--	[42]
Pt/CNT	H, Cl	>98.7/99.8/--	--	[49, 50]
Pt/RGO	H, Cl, Me, OH, NH_2, NO_2, COOH, COMe, OMe, $N(Me)_2$	--/--/>87	--	[51, 52]
Pt/CNF-P-*n*-octylamine	Cl, Br, I, OH, OMe, $CONH_2$, CO_2Et, COMe, OBn, CH(OH)Me, C=C, CN	--/--/>64	--	[13]
Pt/N-CNF-H	Cl, Br, I, OMe, CO_2Et, COMe, CN	--/--/>97	--	[44]
5% Pt-0.5% V/AC-4	Cl	--/--/99.4	--	[43]
Pt(S)/C	Cl, Br, I	--/--/>92	--	[45]
$Pt/Co(OH)_2$	Cl, Br, I	>99.6/100/--	--	[53]
PEG-Pt	Cl	99.7/100/--	1.04/--	[54]
Mw-Pt/Ti-HMS	H	--	--/0.19	[55, 56]
Pt/montmorillonite	H	>82/100/--	17.9/--	[57]
Pt@Si_{c6}	Cl, OH, CO_2Et, COMe, $CONH_2$, OBn	--/--/>97	--	[58]
Pt/PICP	H, F, Cl, Br, I, CHO, CN, NH_2, OH, COMe, Me	--/--/>92	--	[59]
Schiff base Pt(II) -montmorillonite	H, Cl, OMe, COMe, NO_2	>95/>93/--	--	[60]
Pt(II)-MCM-41	Cl	>70/100/--	--	[61]

[a] TOF: turnover frequency, A: average rate, R: real rate.

SMSI was also proposed by Corma et al., they found after the 0.2% Pt/TiO_2 catalyst was reduced at high temperature (723 K), the suboxide TiO_x species preferentially covered Pt crystal terraces and the relative number of metal sites in corners and axes increased, resulting in a preferential adsorption of nitro group to yield *m*-aminostyrene with a good selectivity (up to 99.6% at 96% conversion) during the hydrogenation of *m*-nitrostyrene [16]. The SMSI also occurred between the TiO_2 support and other metals such as Au, Ni, and Ru, which were also effective for the chemoselective reduction of *m*-nitrostyrene, *p*-iodonitrobenzene (*p*-INB), *p*-nitrobenzaldehyde, and *p*-nitrobenzonitrile [16]. Zhou et al. investigated the catalytic performance of γ-Al_2O_3, ZrO_2 and TiO_2 supported Pt catalysts in the hydrogenation of *p*-CNB. The hydrogenation rate and the selectivity to *p*-chloroaniline (*p*-CAN) over Pt/TiO_2 were higher than those over Pt/γ-Al_2O_3 and Pt/ZrO_2 catalysts. They supposed that the SMSI state in the 0.3% Pt/TiO_2 catalyst, which was calcined at high temperature (673 K), would be responsible for the better catalytic properties of Pt/TiO_2 [35]. Liu et al. prepared Pt/TiO_2/RGO (reduced graphite oxide) via immobilizing Pt particles onto the TiO_2/RGO composite. Compared with Pt/TiO_2 or Pt/RGO, the Pt particles on Pt/TiO_2/RGO displayed the most electron-deficient state due to the simultaneous interactions with TiO_2 and RGO, resulting in the production of more hydride ions and further to enhance the catalytic properties of the Pt/TiO_2/RGO catalyst [36].

Yan et al. reported that $Pt/Fe_2O_3/C$ exhibited extremely superior activity, selectivity and stability for the selective hydrogenation of *o*-CNB to *o*-CAN, and the excellent catalytic performance derived from the transfer of electron between highly dispersed Pt nanoparticles and Fe_2O_3 support [37]. Wang et al. assembled a Pt/γ-Fe_2O_3 nanocomposite catalyst, which exhibited excellent catalytic properties for the selective hydrogenation of halonitrobenzenes (HNBs) (*o*-CNB, bromonitrobenzenes (BNBs), *m*-, *p*-INB, 2,4-dinitrochloro-benzene (2,4-DNCB)) in methanol. High selectivity (>99.8%) to haloanilines (HANs) was obtained at 100% conversion of HNBs and the dehalogenation of HANs over this nanocomposite catalyst was fully inhibited even at complete conversion. The Pt/γ-Fe_2O_3 catalyst also exhibited high stability, and its ferromagnetic property provided a convenient separation of catalyst from the reaction system in an applied magnetic field [14, 15, 38]. The same authors also prepared Pt/Fe_3O_4 nanocomposite catalysts with different manners for the selective hydrogenation of HNBs (*o*-, *m*-CNB, *o*-BNB). The Pt/Fe_3O_4 catalyst was less active than the Pt/γ-Fe_2O_3 nanocomposite because Pt particles were much larger on the Pt/Fe_3O_4 nanocomposite (d_{av} 12.5 nm) compared to those on the Pt/γ-Fe_2O_3 catalyst (d_{av} 2.6 nm). However, there was no intrinsic difference in the specific rate (turnover frequency over per surface of Pt atom) between Pt/Fe_3O_4 catalyst (0.10 s^{-1}) and Pt/γ-Fe_2O_3 nanocomposite (0.09 s^{-1}), and the similar catalytic selectivity to HANs was obtained over them [15]. FTIR-CO probe was used to study the interaction between the Pt nanoparticles and iron oxides [14], the Pt/γ-Fe_2O_3 exhibited a strong peak centered at 2070 cm^{-1} (CO adsorbed on the on-top site) and a broad band in the range of 1730–1870 cm^{-1} (CO adsorbed on the bridge and 3-fold sites). However, these typical signals were not detected on the Pt/Fe_3O_4 sample, suggesting that the chemisorption of CO on Pt/Fe_3O_4 is very weak. After Pt/γ-Fe_2O_3 was treated with hydrogen at room temperature (formed Pt/γ-Fe_2O_3-PR), the typical signals of CO adsorbed were not detected either due to the partially reduction of γ-Fe_2O_3 to Fe_3O_4. Weak CO adsorption on Pt/γ-Fe_2O_3-PR implies less back-donation from Pt to CO, suggesting that the Pt particles are in a state of electron-deficient, which was confirmed by XPS measurements [14]. In addition, it was also reported that improvements in the activity and selectivity to HANs may be attributed to the interaction and charge transfer between metal particles and supports in Pt/γ-$Zr(PO_4)(NH_4HPO_4)$ (Pt/γ-ZrP) [39], Pt/HCl-acidified attapulgite (Pt/H^+-Atta) [40], Pt/titania-incorporated silica (Pt/TiO_2-SiO_2) [41], and Pt/MWNT [42].

In early years, intensive efforts have been made to suppress the dehalogenation in the hydrogenation of HNBs by introducing additives such as amine, magnesium oxide, calcium hydroxide, amidine derivatives, triphenylphosphite, phosphorous acid and their derivatives, or by using partly poisoned catalysts. Most of these works have been reviewed by Stratz [63] which will not be discussed here. Recently, Motoyama et al. reported that complete suppression of dechlorination was achieved during the hydrogenation of *p*-CNB with Pt/CNF-P by adding *n*-octylamine, and the yield of *p*-CAN increased from 96.3% to >99.9%. The catalytic hydrogenation of other HNBs (*o*-, *m*-CNB, *p*-BNB, *p*-INB) was also achieved with high yield (>96%) with the Pt/CNF-P-*n*-octylamine system [13]. Thereafter, they synthesized Pt nanoparticles supported on nitrogen-doped (5 wt%) carbon nanofibers (Pt/N-CNF-H), however the reaction rate over Pt/N-CNF-H was reduced to approximately one ninth of that over Pt/CNF-H. Nitrogen atoms in the honeycomb lattice acted as a catalyst poison in the hydrogenation of aromatic nitro compounds, but the catalytic performance can be controlled by an appropriate choice of the Pt/N ratio. The hydrogenation of aromatic nitro compounds over 3% Pt/N-CNF-H afforded the corresponding aromatic amines in high yield (>97%) without reduction of other reducible functional groups (Cl, Br, I, C=O, C=C, CN), whereas reactions with 1% Pt/N-CNF-H gave the intermediate *N*-

aryl hydroxylamines with high yield (>90%) [44]. It has been reported that the performance of Pt/AC could be modified by addition of vanadium salt into the reaction system [5], but it is difficult to separate vanadium salt from the solvent and aromatic amines. Möbus et al. modified Pt/AC with vanadium compounds during the catalyst preparation, both the activity and the yield of desired *o*-CAN increased [43]. Kasparian et al. used a commercially available sulfided platinum catalyst (Pt(S)/C) to catalyze the hydrogenation of nitro groups in the presence of activated heteroaryl halides, in which the dehalogenation was effectively suppressed [45]. Cheng et al. reported that when the $Co(OH)_2$ and 1% Pt/C were used together, the deiodogenation was inhibited and the selectivity to *m*-IAN increased from 91.8% to 97.3% at the complete conversion of *m*-INB. Then they prepared Pt/$Co(OH)_2$ catalyst, it gave 99.6% selectivity to HANs at complete conversion for the hydrogenation of CNB, *p*-BNB and *m*-INB, in which $Co(OH)_2$ acted as both support and dehalogenation inhibitor [53].

2.2. Pd Catalysts

Pd catalyst is widely studied for the catalytic hydrogenolysis of organohalides, which shows a higher catalytic activity for the hydrogenolysis of carbon-halogen bond in the selective hydrogenation of HNBs as compared with Pt, Ru and Rh catalysts. However, Pd-based catalysts exhibit good performance in suppressing the accumulation of harmful intermediates in the hydrogenation of aromatic nitro compounds. Thereby, Pd-based catalysts had been extensively used, and some strategies have been developed for improving the selectivity to HANs in the hydrogenation of HNBs such as introducing the additives to the reaction system [13, 89, 96] or to the catalyst [90], applying partly poisoned catalyst [93], modulating the interaction between support and active species [84], alloying Pd with other metals [96, 97] (in part 2.8), and conducting the reaction in $scCO_2$ [21] (in part 3).

Li et al. prepared a series of PVP-Pd/Al_2O_3 catalysts modified by metal ions (Li^+, Ca^{2+}, Co^{2+}, Mg^{2+}, Fe^{3+}, Fe^{2+}, Ni^{2+}, Sn^{2+} and Sn^{4+}), and Sn^{4+} exhibited the highest promoting effect on activity and selectivity. The selectivity to *p*-CAN markedly increased to 96.6% at *p*-CNB conversion close to 100% with introducing a soluble salt of Sn^{4+} (Sn^{4+}/Pd molar ratio of 1:1) into the reaction system. Increasing the dosage of Sn^{4+} could further improve the selectivity to *p*-CAN, and the dechlorination was completely blocked at a molar ratio of Sn^{4+}/Pd reached 5 [89].

However, the presence of additives brought the difficulty for the product separation and purification. Thereafter, Vishwanathan et al. prepared a series of alkali metals (Li, Na, K and Cs) promoted alumina-supported palladium catalysts for the gas phase hydrogenation of *o*-CNB at 523 K.

Table 2. Hydrogenation of aromatic nitro compounds with Pd-based catalysts

Catalyst	Substituent	Selectivity/ conversion/ yield (%)	TOF A/R [a] (s^{-1})	Ref.
Pd/H_2N-SiO_2/Fe_2O_3	H	--/--/99	0.185/--	[64]
Pd/Al_2O_3/SiO_2/Fe_3O_4	H	>99.9/100/--	--	[65]
Pd/NH_2-Fe_3O_4	H, Cl, Br, Me, OH, COMe	--/--/>92	0.023/--	[66]
Pd/□-Fe_2O_3-PR	Cl	--/--/99.2	0.86/--	[67]
Pd/Fe_3O_4	H, Me, OMe, NO_2, NH_2, COMe, OH, Cl	--/--/>81	0.035/--	[68]
Pd/PVP, PVA, PAA-SiO_2	H	--	--	[69]
Pd/HSO_3-PS-DVB	Cl	>92/>95/--	--/0.16	[70]
Pd/PU	H, Me, NH_2, NO_2, Cl	--/>99/--	--	[71]
Pd/PVC	H, NH_2, NO_2, OH, COOH	--/--/>92	--	[72]
Pd/EGCG-CF	H, Me, OH, NH_2, Cl	--/--/>31	>0.023/--	[73]
SiO_2@Pd-PAMAM	Cl	96/100/--	0.344/--	[74]
PS-Pd-Meimz	H, NH_2, OH, Me, COOH, NO_2	--/>53/--	--	[75]
VPy-NVP-Pd	H, Cl, OH, OMe, NH_2	--/--/>92	--	[76]
PVPA-Pd	H, OH, CHO, OMe, NH_2	--/--/>60	--	[77]
PVP-Pd-V(C_6H_5)$_2$	NO_2	--/--/>80	--	[78]
PEG-Pd	H	100/100/--	0.023/--	[79]
Pd-HTA-OAPE-5	Cl	<30/100/--	--	[80]
Pd/HT	H	98/97/--	--/0.6-0.8	[81, 82]
NAP-MgO-Pd	H, Me, Cl, Br, I, F, NO_2, OH, CH_2OH, C=O	--/--/>90	--	[83]
Pd/TiO_2-NT	Cl	86/100/--	--	[84]
Pd/MCM-41	Cl	--/--/>80	1.1-3.9/--	[85]

Catalyst	Substituent	Selectivity/ conversion/ yield (%)	TOF A/R a (s-1)	Ref.
Pd/AlO(OH)	F, NH_2, COMe, OH	--/--/>92	--	[86]
Silia *Cat* Pd^0	H, OH,COOH, $CONH_2$, COOMe, Cl	--/100/--	--	[87]
SiO_2@Pd@ZrO_2	H	100/93/--	0.33/--	[88]
Sn^{4+}-PVP-Pd/Al_2O_3	Cl	99.2/100/--	--	[89]
Na-Pd/Al_2O_3	Cl	>95/73/--	--	[90]
Pd-PPh_3/diamond	H, OH, Cl, COOH, NO_2	--/--/97-99	--	[91]
Pd/C-HCl	$CH_2N(R_1)R_2$	--/--/>96	--	[92]
Pd/C[Ph_2S]	Cl, COMe, CO_2Bn	--/--/>91	--	[93]
LPdW/MCM-41	OH, OMe, $CONH_2$, $COOC_2H_5$	--/--/>96	--	[94]
Pd/C(en)	$CO_2(CH_2)_2OBn$	--/--/98	--	[95]
n-octylamine-Pd/CNF-P	Cl	--/--/99.9	--/0.046	[13]

[a] A: average rate, R: real rate.

The catalytic activity decreased with increasing the ionic radius of alkali metal ions, and the stability of Na-Pd/Al_2O_3 was better than that of Li-Pd/Al_2O_3. The Na content had large influence on the activity and selectivity of Na-Pd/Al_2O_3 and the optimum Na content was 0.5-3%. A considerable decrease in conversion and selectivity presented at the content of Na above 5%. The authors proposed that a strong interaction occurred between the alkali ion (M^{+1}) and oxygen of -NO_2 group. This increased the polarity of N=O bond and made it easier to be attacked by the dissociated hydrogen. Moreover, the Na-Pd/Al_2O_3 showed better resistance towards coke formation, which might be attributed to the variation in acid-base distribution on the catalyst in the presence of Na_2O [90]. Kantam et al. prepared nanocrystalline magnesium oxide-stabilized Pd heterogeneous catalyst (NAP-MgO-Pd) for the hydrogenation of aromatic and aliphatic nitro compounds to the corresponding amines. During the reduction of *p*-CNB and *p*-nitrofluorobenzene, no dehalogenated products were detected; by contrast, trace amounts of the dehalogenated products were formed in the reduction of *p*-INB and *p*-BNB. The NAP-MgO-Pd catalyst could be recovered and reused up to ten cycles without further activation. The presence of edge-corner and other defect sites allowed the nanostructured MgO to possess the reactive surface sites, and the coordination of O^{2-} and Mg^{2+} could readily accept incoming reagents with Lewis acid or Lewis base character [83]. SMSI plays an important role in improving the activity and selectivity of supported Pd catalysts for the hydrogenation of aromatic nitro compounds. Coville et al. prepared Pd/TiO_2-C (Degussa P-25) and Pd/TiO_2-M (the nanotubes synthesized from Degussa P-25); over Pd/TiO_2-M catalyst, both the conversion of *o*-CNB (100%) and the selectivity to *o*-CAN (84%) was higher than those of Pd/TiO_2-C catalyst, but the selectivity to *o*-CAN was still lower and the dechorination occurred even at the optimized reaction conditions [84]. Wang et al. reported that the electronic interaction between Pd and γ-Fe_2O_3 played an important role in enhancing the selectivity to *p*-CAN over the Pd/γ-Fe_2O_3-PR and inhibiting the dechorination reaction in the hydrogenation of *p*-CNB [67] as described above for the Pt/γ-Fe_2O_3-PR catalyst [14, 15, 38].

2.3. Ru Catalysts

Among the Pt-, Pd-, Ru-, Rh-, Ir- and Au-based catalysts, Ru is cheaper than the other noble metals and more selective for the hydrogenation of aromatic nitro compounds to aromatic amines, but its catalytic activity is generally low. Therefore, many approaches have been developed to enhance the activity of Ru-based catalyst [16, 98-107], and some achievements have been obtained, which will be related below.

Table 3. Hydrogenation of aromatic nitro compounds with Ru-based catalysts

Catalyst	Substituent	Selectivity/ conversion/ yield (%)	TOF A/R [a] (h^{-1})	Ref.
Ru/SBA-15	H	99.9/96.5/--	--	[98]
Ru/C	NO_2	>97/>95/--	--	[108]
PVP-Ru/Al_2O_3	NO_2	100/93.9/--	--	[100]
PVP-Ru/Al_2O_3-Sn^{4+}	NO_2	97.9/100/--	--	[100]
Ru-Pb, Sn, Ge/γ-Al_2O_3	Cl	--/<70/--	--/17.1	[107]
Ru/SiO_2	Cl	98.6/99.9/--	115/--	[105]
Ru/SnO_2	Cl	>99.9/100/--	155/--	[101]
Ru/CNT, CNF	Cl	92-94/100/--	--/1900	[103]
Ru/MgF_2	Cl	>97/--/--	--	[104, 106]
Ru-V/MgF_2	Cl	>97/>98/--	--	[104]
Ru/TiO_2	C=C, CHO, CN, I	>85/>90/--	--	[16]

[a] A: average rate, R: real rate.

The catalytic performances can be altered significantly by controlling the metal particle dispersion and metal–support interactions which can modify the electronic state and/or morphology of Ru crystals, and so improve the catalytic properties. Wang et al. reported that the catalytic activity of the Ru/SnO_2 nanocomposite catalyst was much higher than the PVP–Ru nanoclusters and Ru/SnO_2 prepared by the traditional impregnation method. The Ru/SnO_2 nanocomposite catalyst can be reused for four times without loss of activity and the selectivity to *o*-CAN kept above 99.8%. The outstanding catalytic selectivity over the Ru/SnO_2 nanocomposite catalyst was due to a relative strong adsorption of *o*-CNB on the catalyst, the oxygen vacancies or coordinatively unsaturated Sn^{4+} or Sn^{2+} species of SnO_2 nanoparticles interacted with the Ru particles and thus its activity was accelerated for the hydrogenation of $-NO_2$ group, and suppressed for the dehalogenation of *o*-CAN [101, 102]. Serp et al. used carbon nanotubes (CNTs) and carbon nanofibers (CNFs) as catalysis support to prepared Ru/CNT and Ru/CNF catalysts, they found the catalytic activities of Ru/CNT and Ru/CNF (TOF: 1900 h^{-1}) catalysts in the hydrogenation of *p*-CNB were much higher than those of the commercial Ru/Al_2O_3 and Ru/C catalysts and other Ru-containing catalysts (TOF: 30-400 h^{-1}); however, the selectivity to *p*-CAN were only 92-94% and dechlorogenation occurred over them. The increased activity over Ru/CNT and Ru/CNF was attributed to the following reasons: (a) The supports of CNF and CNT could avoid the diffusion limitations and maintain a high hydrogen flux on the catalyst surface; (b) The electronic interaction generated electron-deficient Ru centers resulting in an increase of the catalytic activity [103]. Pietrowski et al. [104, 106] reported that Ru/MgF_2 exhibited high activity and 100% selectivity to *o*-CAN for the hydrogenation of *o*-CNB in liquid phase. The activity of Ru/MgF_2 was further enhanced by the addition of vanadium. The VO_x species, formed after reduction with hydrogen at 673 K, could cover the surface of Ru crystallites and modify their electronic properties and generate new active Ru^{n+} sites, resulting in a much more active Ru-V/MgF_2 catalyst [104].

Xu et al. reported that the selective hydrogenation of *p*-CNB to *p*-CAN over Ru/SiO_2 catalyst was promoted remarkably by simply adding 30% water (in volume) into reaction solution; the selectivity increased from 80.4% to 98.6% and the TON increased from 0.89×10^{-2} to 3.20×10^{-2} s^{-1} [105]. The promoting effect of water was also found over Fe/SiO_2, Co/SiO_2, Ni/SiO_2, Cu/SiO_2, Ag/SiO_2 and Ru/MgF_2 catalysts [105, 106]. Zhao et al. reported that the polylvinylpyrrolidone stabilized Ru/Al_2O_3 (PVP-Ru/Al_2O_3) showed higher conversion (93.9%) and higher stability in the selective hydrogenation of *m*-

dinitrobenzene (*m*-DNB) to *m*-nitroaniline (*m*-NAN) compared to Ru/Al_2O_3 catalyst (73.6%). Furthermore, the solvent and metal cation additive presented a large influence on the hydrogenation of *m*-DNB; protic solvents were more favorable for the reaction than non-protic solvents. The additive of metal cations, such as Co^{2+}, Ni^{2+}, Li^{+} and Sn^{4+}, could markedly enhance the reaction rate. Maximum yield of *m*-NAN (97.9%) was obtained in the presence of a low concentration of the Sn^{4+} in methanol [100].

For the hydrogenation of *p*-CNB, Figueras et al. found that the doping of Pb, Sn and Ge to the $Ru/\gamma\text{-}Al_2O_3$ catalyst could significantly improve the catalytic activity. The effect of Ru dispersion on the catalytic activity was also marked and the specific activity (TOF) increased with decreasing of dispersion [107].

2.4. Au Catalyst

Au element is traditionally considered having poor performance in catalysis, while this situation changed in the last decade when several cases of gold-catalyzed hydrogenation of C=C and C=O groups were reported, and gold exhibits some selectivity for the hydrogenation of C=O groups of α, β-unsaturated aldehydes, producing allyl-type unsaturated alcohols [109, 110]. In 2006, Corma and Serna [10, 123] as well as Chen et al. [114] foremost explored gold catalyst for the reduction of nitro compounds, and supported gold exhibited high chemoselectivities for the hydrogenation of aromatic nitro compounds containing Cl, C=C, C=O, or CN substituents. Thereafter, a series of supported gold catalysts were prepared for the hydrogenation of aromatic nitro compounds with functional group (Cl, I, C=C, C=O, CN, NO_2); the influences of particle size and support, and the reaction pathway were studied.

Corma and Serna prepared 1.5% Au/TiO_2 and 4.5% Au/Fe_2O_3 catalysts with a mean particle diameter of 3.7–3.8 nm which exhibited high chemoselectivity for the hydrogenation of nitro group in *m*-nitrostyrene (S 95.9%, C 98.5%), *p*-nitrobenzaldehyde (S 96.8%, C 99.0%), and *p*-nitrobenzonitrile (S 97.2%, C 99.0%) [10]. The high chemoselectivity of Au/TiO_2 is due not only to an intrinsic higher rate for the hydrogenation of the nitro versus the olefinic group on the gold catalyst, but also to the very favorable adsorption through the nitro when both groups are competing [111, 124]. Over the most selective Au/TiO_2 catalyst, H_2 is dissociated on gold particles while nitrostyrene is weakly adsorbed on metallic Au and more strongly on highly uncoordinated Au atoms. Nitro groups relatively weakly bound to the TiO_2 surface but more strongly to the metal–support interface. It was confirmed by the density functional (DFT) calculations that an interaction existed between NO_2 group and highly uncoordinated Au atoms at the interface of the gold particle and TiO_2 support, which could explain the accelerated reactivity of the nitro group [111].

Table 4. Hydrogenation of aromatic nitro compounds with Au-based catalysts

Catalyst	Substituent	Selectivity/ conversion (%)	Ref.
Au/TiO_2, Fe_2O_3	CN, CHO, $CONH_2$, C=C	>95/>98	[10, 111]
Au/TiO_2	Cl	100/--	[112]
Au/Fe_2O_3, Fe_3O_4	Cl, NO_2	--	[113]
Au/SiO_2	Cl, OC_2H_5, COMe, CH_3, OH	>95/100	[114]
Au/SiO_2-org	H, CH_3, Cl, NO_2	>95/>99	[115]
Au/Al_2O_3	C=C, COMe, $CONH_2$	>89/>95	[116]
Au/Al_2O_3	Cl	100/--	[112, 117]
Au/ZrO_2	Cl	>99.4/100	[118]
$Au/Fe(OH)_x$	Cl, CHO, COMe, CN	>99/--	[119]
Au/Mo_2N	Cl	100/--	[120]
Au/Mo_2C, $Au/Mo_2C/Al_2O_3$	Cl, NO_2	--	[121]
Au (III) complex	CH_3, Cl, Br, NH_2, OCH_3, C=O, C=C	--/>99	[122]

Gold particles supported on TiO_2 or Fe_2O_3 had a high chemoselectivity in the reduction of various functionalized aromatic nitro compounds without PHA accumulation or by-product formation [10, 111]. However, the formation of NSB and PHA was observed over Au/TiO_2 based on a combination of *in situ* IR spectroscopy with macrokinetic experiments (0.1 mbar NB, 6.3 mbar H_2, 393 K), indicating NSB and PHA adsorbed strongly on the catalyst surface and reacted out before their desorption during the batch-reaction study. Moreover, the amount of PHA was much higher in the starting reactant of NB than that of NSB, it suggested that a direct route from NB to PHA also contributes to the production of AN, and the reduction of PHA appears to be the rate-determining step [4]. This was in agreement with the results reported by Gelder et al., who showed that it was possible to reduce NB to PHA directly with a Pd catalyst on the basis of isotopic results [2, 125]. van Bokhoven et al. [126] monitored the liquid phase and the solid/liquid catalyst interface by *in situ* ATR-FTIR to identify the species adsorbed on Au/TiO_2

and those in the liquid phase during the three-phase hydrogenation of NB (10.4 mmol NB, 1.0 MPa H_2, 25 ml toluene, 393 K). They suggested the hydrogenation of NB over Au/TiO_2 proceeds through the direct route, which was in agreement with that presented by Corma [4]. However, the hydrogenation of NB was reported to proceed through the condensation route over the Au/CeO_2 catalyst, because the rate of hydrogenation was considerably lower over Au/CeO_2. As a result, NSB accumulated and formed condensation intermediates during the reaction. In addition, accumulated NSB could also come from the decomposition of PHA over the CeO_2 support, which was especially fast. So the support has a direct impact on the reaction mechanism and significantly changes the reaction route [127].

Shimizu et al. prepared gold catalysts such as Au/Al_2O_3, Au/SiO_2, Au/MgO, Au/montmorillonite, Au/C and Au/TiO_2 for the hydrogenation of nitrostyrene, and found that the activity depended strongly on metal particle size and the properties of support. The small Au NPs (2.5 nm) on the acid-base bifunctional support (Al_2O_3) gave the highest intrinsic activity. Cooperation of the acid-base pair sites on Al_2O_3 and the coordinatively unsaturated Au atoms on the Au NPs are responsible for the H_2 dissociation to yield H^+/H^- pair at metal/support interface. High chemoselectivity can be attributed to a preferential transfer of H^+/H^- to the polar bonds of nitro group [116]. Keane et al. prepared Au supported on TiO_2, Fe_2O_3, CeO_2 and Al_2O_3 catalysts for the hydrogenation of *m*-DNB to *m*-NAN or *m*-phenylenediamine (*m*-PDM), and they found that Au/TiO_2 and Au/Fe_2O_3 promoted the exclusive hydrogenation of *m*-DNB to *m*-NAN, Au/CeO_2 delivered *m*-PDM as the sole product and Au/Al_2O_3 generated a mixture of *m*-NAN and *m*-PDM at about 10% conversion of *m*-DNB. Furthermore, an increase in specific hydrogenation rate was observed with a decrease in mean Au size from 9 to 3 nm, and a lower specific rate recorded for Au particles <2 nm which was attributed to a quantum size effect. The support plays an important role for the selective hydrogenation of aromatic nitro compounds with substituent groups as described above [128]. Keane et al. also confirmed that Au-based catalysts achieved very high chemoselectivities for the hydrogenation of a variety of HNBs even under continuous gas-phase conditions albeit at incomplete conversions [112, 113, 117, 120, 121]. Au/Al_2O_3 catalyst achieved 100% selectivity for the hydrogenation of *o*-, *m*-, and *p*-CNB and *p*-BNB, as well as 2,4-, 3,4- and 3,5-dichloronitrobenzene [117]. Au/TiO_2 catalysts had the same selectivity as Au/Al_2O_3 for the hydrogenation of CNBs but the activity was higher by a factor of four. The higher activity of Au/TiO_2 might be attributed to the combination of smaller Au particle size with higher number of defect sites for hydrogen activation and possible *p*-CNB activation through interaction with TiO_2 surface oxygen vacancies [112]. Keane et al. also reported that the nature of the support (Al_2O_3, CeO_2, Fe_2O_3, TiO_2 (with anatase:rutile ratio)) did not impact directly on the rate of nitro-group reduction, which is governed by Au particle size [129, 130].

2.5. Ag Catalysts

Supported silver catalysts have long been extensively studied as oxidizing catalysts and have been applied industrially to the epoxidation of ethylene [131, 132]. However, the catalytic reduction properties of silver have not drawn much attention, especially for the catalytic reduction of fine chemicals. Qiu et al. reported that Ag/SiO_2 catalyst, prepared by an *in situ* reduction method, was highly effective and recyclable for the selective hydrogenation of *p*-nitrotoluene, *o*-nitrophenol, and a range of CNBs, giving 100% selectivity to the corresponding aromatic amines [133]. Xu et al. designed and prepared Ag/C_{60} nanocomposite catalyst by reducing $AgNO_3$ with C_{60} monoanion in one-pot reaction. The Ag/C_{60} catalyst exhibited excellent activity and selectivity in the hydrogenation of CNBs and nitrobenzaldehydes to the corresponding aromatic amines. No hydroxylamine intermediate was detected in the reaction process. The charge transfer at the interface between the metal center and the support played a crucial role for excellent catalytic performance [134]. Shimizu et al. studied the influence of silver metal particle size and support oxides in the chemoselective hydrogenation of aromatic nitro compounds. They found that smaller Ag nanoparticles on the acid-base bifunctional support (Al_2O_3) gave higher intrinsic activity, and good to excellent selectivities (86-97%) at 100% conversion were obtained in hydrogenation of a series of aromatic nitro compounds with C=C, COMe, $CONH_2$ and CN groups. The hydrogenation of NB occurs through the consecutive route via the PHA intermediate, and silver cluster plays an important role in the H_2 activation step and the cleavage of the H-H bond is the rate-determining step. Cooperation of the acid–base pair site on Al_2O_3 and the coordinatively unsaturated Ag sites on the Ag cluster is responsible for the rate-limiting H_2 dissociation to yield H^+/H^- pair at metal/support interface, while the basic site on Al_2O_3 acts as an adsorption site of aromatic nitro compounds. High chemoselectivity could be attributed to a preferential transfer of H^+/H^- to the polar bonds of nitro group [135]. Important roles of the interface between the support oxide and the metal surface atoms have been proposed in the literature, but most of them have been focused on the reducible character of semiconductor oxides such as TiO_2 [34, 111]. Kaneda et al. reported that silver nanoparticles on a basic support of hydrotalcite (Ag/HT) could catalyze the chemoselective reductions of nitrostyrene to the aminostyrene selectively with alcohols or CO/H_2O as reducing reagent while retaining the reducible C=C bond [136]. However, the use of H_2 instead of alcohols or CO/H_2O caused reductions of both the polar groups (nitro) and the nonpolar C=C bond. This nonselective reduction was due to the formation of nonpolar hydrogen species through the homolytic cleavage of H_2 at the AgNPs surface, which was active for the reduction of C=C bond. So they envisioned that AgNPs covered with a basic material (BM), the core–shell nanocomposite AgNPs@BM, would be a reasonable structure for performing the above complete chemoselective reduction. The AgNPs@BM structure can maximize the interface area of the AgNPs–BM, while minimizing the area of the bare AgNPs. This property would enable the exclusive formation of the heterolytically cleaved hydrogen species through a concerted effect between AgNPs and basic sites of BM that suppressed the unfavorable formation of homolytically cleaved hydrogen species on the bare AgNPs. The resulting Ag hydride and proton species would lead to complete chemoselective reduction of polar functionalities while retaining the C=C bond. So Kaneda et al. synthesized a novel core–shell AgNPs–CeO_2 nanocomposite (AgNPs@CeO_2) involving core AgNPs 10 nm in diameter and a shell assembled with spherical CeO_2 NPs 3–5 nm in diameter. The shell had nanospaces between the CeO_2 NPs that permit the access of reactants to the active center of the AgNPs. The AgNPs@CeO_2 catalyst showed complete chemoselectivity for the reduction of nitrostyrene to aminostyrene by using H_2 while retaining the reducible C=C bond. The advantages of the core–shell structure are shown in the remarkable improvement of the chemoselectivity compared with AgNPs supported on CeO_2 (AgNPs/CeO_2). Moreover, AgNPs@CeO_2 was easily separable from the reaction mixture and was reusable without loss of catalytic activity or selectivity [11].

2.6. Ir and Rh Catalysts

Ir supported catalysts are rarely used in the hydrogenation of HNBs, because the hydrogenation catalyzed by Ir is accompanied by the hydrogenolysis of carbon–halogen bond. The catalytic property of Ir could be modified with the second metal component, such as the dehalogenation can be usually kept below 0.1% if Fe and Cu were used as doping elements, and the using of Ir-Fe-Co and Ir-Fe-Mn trimetallic catalysts not only significantly improved the selectivity of the hydrogenation, but also shorten the reaction time in the hydrogenation of *o*-CNB. The Ir-Fe-Cu trimetallic catalyst also exhibited high yield (99.3%) in the hydrogenation of 2,4,5-trichloro-nitrobenzene to 2,4,5-trichloroaniline [137]. Li et al. reported that the hydrous zirconia supported Ir nanoparticles ($Ir/ZrO_2{\cdot}xH_2O$) exhibited an excellent conversion (100%), selectivity (>99.9%) and stability under room temperature and atmospheric pressure for the hydrogenation of HNBs in the mixture ethanol and water. It is suggested that the hydroxyl groups on the surface of the catalyst form the hydrogen bond with N=O bond of CNB and cause the polarization of N=O bond, so the activated hydrogen on the surface of Ir particles is more easily to attack the polarized N=O bond. The hydrogen bond between the substrate and water also plays an important role to the activation of substrate molecule [138].

The catalytic properties of Rh/C are essentially similar to those of Pt/C, as exhibited for the hydrogenation of BNBs [139]. However, Rh-based catalyst has received less attention because the price of Rh is much higher than that of Pt. In addition, dehalogenation has been shown to occur over the catalysts of Rh on activated carbon in the hydrogenation of *p*-CNB [140].

2.7. Non-Noble Metal Ni and Co Catalysts

Non-noble metals such as Ni and Co are abundant and cheap, and are of great significance in catalysis. However, the activity of supported Ni or Co catalysts is usually lower compared with that of Pt and Pd. Recently, many efforts have been made to enhance the hydrogenation activity of Ni and Co catalysts [141-155], and will be described below.

The acidity of support significantly affects the catalytic activity of Ni catalyst. Yin et al. reported that Ni/kieselguhr with the weakest acidity had much higher TOF than Ni/Al_2O_3 and Ni/TiO_2 for the hydrogenation of *p*-nitrophenol to *p*-aminophenol [141]. Ding et al. reported that addition of K_2O could improve the level of *p*-nitrophenol adsorption over Ni/AC catalyst, and the reaction efficiency was enhanced consequently [142]. However, Nieto-Márquez et al. showed that the activity of Ni/AC was insensitive to Ni particle size and exhibited a proportional increase with increasing surface acidity (mainly carboxylic anhydrides) for the gas-phase hydrogenation of NB to AN. These authors attributed this response to an enhanced activation of NB at surface acid sites [143]. SMSI also plays an important role in enhancing the activity and selectivity of supported Ni catalysts for the hydrogenation of aromatic nitro compounds.

Table 5. Hydrogenation of aromatic nitro compounds with Ni- and Co-based catalysts

Catalyst	Substituent	Selectivity / conversion/ yield (%)	TOF A/R [a] (h^{-1})	Ref.
Ni/kieselguhr	OH	>99/99.4/--	--/40	[141]
Ni/AC, Ni-K_2O/AC	OH	99.2/97.7/--	--	[142]
Ni/AC, CNS, CNF	H	100/--/--	82/--	[143]
Ni/TiO_2	Cl	99.6/99.9/--	49.2/--	[144]
Ni/TiO_2-IM	Cl	>98/>99/--	--/8640	[145]
Ni/TiO_2 (sol-gel)	Cl	>99/>99/--	--	[146]
Ni/TiO_2	OH	100/100/--	--	[150]
Ni/TiO_2	C=C, CHO, CN, I	>88.3/>93/--	--	[16]
NiFC	H, Cl	--/--/>99	50.5/--	[147-149]
Ni/Al_2O_3	H, Cl, NO_2	100/--/--	58/--	[6]
Ni/SiO_2 (commercial)	H	--	--	[156]
Ni/SiO_2 (passivated)	H, Cl, CH_3, CHO, OH, OCH_3 etc.	100/100/--	--	[157]
Ni/$SiO_{2\text{-}EN}$	H	99.2/100/--	--/0.83	[151]
Ni/CNF	Cl	98/99/--	--/504	[152]
Ni/L-HCNT	Cl	98.3/97.1/--	--	[153]
Ni NPs	OH	--/98/--	5.63/--	[155]
Nano-size Ni	OH	--	--	[158]
Nano-size Ni	OH	--	74.8/--	[154]
(Ni/TiO_2)@C	H	94.8/97/--	100.9/--	[159]
Ni/SBA-15, MgO	H	100/>99/--	--	[160]
Ni/K_2O-La_2O_3-SiO_2	NO_2	99.6/99.9/--	--	[161]
Ni/SBA-15	Cl	97.9/97.2/--	--	[162]
Co/SiO_2	H, OH	100/>89.8/--	>17000/--	[163]
Co/C	Cl	97/98/--	--	[164]

[a] A: average rate, R: real rate.

Chen et al. investigated the performance of Ni catalysts supported on SiO_2, ZrO_2, TiO_2 and γ-Al_2O_3 in the hydrogenation of *o*-CNB. Ni/TiO_2 was the best catalyst that gave 99.6% selectivity to *o*-CAN at 99.9% conversion of *o*-CNB, although the amount of H_2 adsorption on Ni/TiO_2 was much lower than that on Ni/ZrO_2 and Ni/SiO_2. The better performance of Ni/TiO_2 was attributed to the strong polarization of N=O bond induced by oxygen vacancies of TiO_x that was generated by reduction at high temperature [144]. The interaction between Ni and TiO_2 is also influenced by the temperature of TiO_2 pretreatment. With increasing the TiO_2 calcination temperature from 773 to 1173 K, the specific surface area and the pore volume of Ni/TiO_2 decreased, the NiO–TiO_2 interaction became weak, and the Ni surface area and the catalytic activity increased initially and then decreased. There was a proper calcination temperature, at which the catalyst showed the best activity [145]. Chen et al. also prepared Ni/TiO_2 catalyst by a sol–gel method and investigated the effect of calcination and reduction temperature on its catalytic performance. The increase of the calcination temperature from 673 to 823 K decreased the surface area and enhanced the interaction between Ni species and support, resulting in the decrease of the available Ni active sites and the catalyst performance. With raising the reduction temperature from 623 to 773 K, the interaction between the Ni species and TiO_x was also enhanced, and the sintering of metallic Ni crystallites became serious. Accordingly, the activity of Ni/TiO_2 decreased [146]. Figueiredo et al. prepared Ni nanoparticles stabilized by filamentous carbon (NiFC) for the hydrogenation of NB and CNB in methanol. The catalyst exhibited excellent performance and produced clean aromatic amines without accumulation of hazardous intermediates or any other side products, which can be attributed to the electron enrichment of the Ni particles [147-149].

Generally, the dispersion of Ni catalysts has some influences on the activity. Hou et al. prepared a Ni/SiO_2 catalyst with an average Ni particle size of 3.7 nm via impregnation of $Ni(en)_3(NO_3)_2$ on a silica gel. The catalyst was more efficient than Raney Ni for the hydrogenation of NB to AN [151]. Ni particles with the size of 5.1-5.7 nm dispersed on carbon nanofibers (CNFs) also showed good activity and selectivity (up to 95%) for the hydrogenation of CNBs to the corresponding aromatic amines [152]. Li et al. prepared uniformly dispersed Ni nanoparticles with an average size of 5.1 nm by reduction of Ni-Al layered double hydroxide precursor supported on nitric acid modified carbon nanotubes (Ni/L-HCNT). The catalytic performance of Ni/L-HCNT for the liquid phase hydrogenation of *o*-CNB to *o*-CAN was better than that of Ni catalyst supported on CNTs prepared by incipient wetness impregnation [153].

Deactivation of Ni catalysts usually occurs in the hydrogenation of aromatic nitro compounds in organic solvents. The deactivation of Ni/TiO_2 and Ni/K_2O-La_2O_3-SiO_2 was mainly attributed to the surface coverage and the pore blockage with bulky molecular species in the hydrogenations of *o*-CNB to *o*-CAN and *m*-dinitrobenzene to *m*-phenylenediamine [146, 161]. The deactivation of nano-sized Ni particles in the hydrogenation of *p*-nitrophenol was found to be dependent on the initial pH of the solution. At pH=1.6, the deactivation was a combination effect of severe agglomeration of Ni particles, Ni dissolution, and the strong adsorption of the complex compound of Ni ion and *p*-nitrophenol on the catalyst surface. At pH>7.6, the formation of nickel hydroxide on the Ni surface was proposed to be the main reason [158]. Rao et al. reported that Ni/MgO suffered a huge loss in activity during the gas phase hydrogenation of NB to AN and the deactivation might be due to the incomplete reduction of Ni–MgO interacted species and due to the release of water during the reaction. In contrast, Ni/SBA-15 catalyst exhibited superior catalytic activity, selectivity, and stability during operation due to the hydrophobic nature of the catalyst (resistant to water generated during the reaction) and the effect of coking is not significant [160]. It was also reported that NiFC [148, 149], Ni/TiO_2 catalysts prepared by electroless plating technique [150], and Ni/L-HCNT [153] showed high stability in the selective hydrogenation of aromatic nitro amines to aromatic amines. Zheng et al. reported that the Ni/SiO_2 catalyst passivated with O_2 can be recycled for five times during the reduction of various aromatic nitro compounds, although its activity was low after passivation [157].

Co was also employed as an active component for the hydrogenation of aromatic nitro compounds. Raja et al. reported that Co nanoparticles supported on the non-ordered mesoporous silica (Co/SiO_2) were cheap, recyclable and industrially viable catalyst for the hydrogenation of a range of aromatic nitro compounds under mild conditions (353–373 K, 2.5 MPa H_2). The Co/SiO_2 exhibited high selectivity (100%) and turnover frequencies (up to 23119 h^{-1} for NB hydrogenation) which was comparable to that of NiPd/SiO_2 [163]. Qiu et al. prepared Co/C microspheres for the hydrogenation of CNBs at 2 MPa and 413 K; the catalyst showed a conversion of ≥97% and a selectivity of ≥97% to the corresponding CANs [164].

2.8. Bimetallic Catalysts

In many cases, bimetallic catalysts exhibited better activity, selectivity and stability as compared with monometallic catalysts. These phenomena may be due to alloying, structure effect, ensemble effect, and electronic effect. Surface distribution and interaction between the two metals may be taken into account for the observed catalytic performance.

Supported Au catalysts exhibited unique chemoselectivity in the hydrogenation of aromatic nitro compounds but the catalytic activity of such catalysts should be further improved, and the application of bimetallic systems (containing Au) has been considered. Keane et al. reported that the chemical incorporation of Pd in Au/Al_2O_3 had an impact on the conversion and product distribution in the hydrogenation of *p*-CNB.

Table 6. Hydrogenation of aromatic nitro compounds with bimetallic catalysts

Catalyst	Substituent	Selectivity/ conversion (%)	TOF A/R [a] (h^{-1})	Ref.
Au-Ni/Al_2O_3	NO_2	--	--	[165-167]
Au-Pd/Al_2O_3	Cl	100/<10	--	[97, 117]
Au-Pt/ MCM-41	CH_3	--/99	--	[168]
1.5%Au-0.01%Pt/TiO_2	C=C	93.4/94.5	550/--	[30]
0.5%Au-0.02%Pt/TiO_2	Cl	100/100	--	[31]
Ni/Pd nanoclusters	H, CH_3, OCH_3, CN etc.	100/100	576/--	[169, 170]
Ni-Pd/SiO_2	H, OH	>97/--	--/>9600	[163]
Cu-Pd/CeO_2	NO_2, C=C, COMe, COOH	>98.5/>96.7	--	[171]
Ni–Pt/C	NO_2	99.2/100	--/200	[32]
Pt-Ni hollow-nanospheres	Cl	99/99	--/2232	[33]
PtM/TiO_2 (M = Cr, Mn, Fe, Co, Ni and Cu)	Cl	>95.7/>99.3	--/>3060	[35, 172]
PtM/ZrO_2 (M = Mn, Fe and Co)	Cl	>95/>99.4	--/>432	[173]
Pt–Pd/PMO-SBA-15	H	>99/100	--	[29]
Pt–Ru/Fe_3O_4/C	Cl	>99.5/100	--	[28]
PVP-Pt-Ru	Cl	>93/>99.4	--/191	[174]
Cu-Ni/CeO_2–SiO_2	Cl	97.8/96.6	--	[175]
Cu-Ni/Al_2O_3	Cl	100/--	--	[176]

[a] A: average rate, R: real rate.

The inclusion of Pd (at Pd/Au≤0.05) resulted in increased activity while retaining exclusivity to *p*-CAN due to Pd–Au surface interaction. At higher Pd contents (Pd/Au>0.05), a further increase in reaction rate was accompanied by the formation of NB as a result of a Pd promoted hydrodechlorination. The physical mixtures of Au/Al_2O_3 and Pd/Al_2O_3 (Pd/Au=0.05) delivered higher reaction rates but the formation of NB and AN was detected. These authors attributed the enhanced and exclusive production of *p*-CAN over $Pd-Au/Al_2O_3$ to a surface Pd-Au synergism [97, 117]. For the hydrogenation of *m*-nitrostyrene to *m*-aminostyrene over Au/TiO_2, it was found that the rate-controlling step was the dissociation of H_2 on gold according to kinetic and isotopic results from Corma et al [30]. They also prepared a bimetallic $Au@Pt/TiO_2$ catalyst with an optimized Pt content, and the presence of Pt on the bimetallic catalyst increased the rate of H_2 dissociation, resulting in almost one order of magnitude active than the monometallic Au/TiO_2, while maintaining high chemoselectivity to *m*-aminostyrene. By means of density functional theory (DFT) calculations, they found that adsorption modes of *m*-nitrostyrene on Au_{13}/TiO_2 and Pt_{13}/TiO_2 are completely different. No matter how small the Pt particles are, the aromatic ring of *m*-nitrostyrene always interacts strongly with Pt atoms. The molecule is preferentially adsorbed on top of the particle and the two groups are simultaneously activated, explaining the lack of selectivity of Pt catalysts. However, when only small amounts of highly dispersed Pt are added to the catalyst, as simulated by the $Au_{12}Pt/TiO_2$ model, the activity toward hydrogenation increases considerably while the properties of the metal-support interface and the mode of adsorption of *m*-nitrostyrene remain similar to those found for Au_{13}/TiO_2, and therefore the chemoselectivity is maintained [177]. Xu et al. [31] and Zhao et al. [178] have also confirmed the activity enhancement of Au nanoparticles by adding a second metal due to increased H_2 dissociation and hydrogen spillover.

Toshima et al. [169, 170] reported that polymer-protected Ni/Pd bimetallic nanoclusters in a certain intermediate composition range exhibited higher catalytic activity than either a Ni or Pd monometallic nanocluster, and at a certain Ni/Pd mole ratio (2/3), the maximum catalytic activity was observed in the hydrogenation of aromatic nitro compounds. The forming of Ni-Pd alloy may be one of reasons for the high catalytic performance [169]. Raja et al. prepared Ni-Pd bimetallic nanoparticles with a Ni/Pd ratio of 1/4 at lower temperature (353-393 K), and supported Ni-Pd nanoparticles on SiO_2, the catalyst showed highly active, selective, and stable for the hydrogenation of NB, *m*-nitrophenol and *m*-nitro-*o*-cresol [163]. Sun et al. prepared $Pt–Ru/Fe_3O_4/C$ as an excellent catalyst for the solvent-free selective hydrogenation of *o*-CNB to *o*-CAN. The interaction between Fe_3O_4 and Pt-Ru nanoparticles resulted in that Fe_3O_4 nanoparticles were electron-enriched, and Pt and Ru were electron-deficient. The high selectivity and activity of Pt–

$Ru/Fe_3O_4/C$ catalyst might be attributed to the electron transfer between these two metals and Fe_3O_4. Meanwhile, Ru as the electron-deficient species on the surface of the catalyst might act as electrophilic or Lewis acid sites for the adsorption of *o*-CNB, and favoring its hydrogenation [28].

Reddy et al. reported that among different catalysts (CeO_2–SiO_2 supported Cu, Cu-Ni, Cu-Co) employed for the hydrogenation of *o*-CNB, the Cu-Ni/CeO_2-SiO_2 one exhibited a high conversion (96.6%) and product selectivity (98%) without fast deactivation. The high activity and stability of bimetallic catalysts was attributed to better dispersion and smaller particles formation as well as high specific surface area of the catalyst due to the addition of Ni to Cu [175]. Kiwi-Minsker et al. reported that the incorporation of Ni to Cu-Al catalyst derived from hydrotalcite resulted in a significant increase in hydrogenation rate, whereas copper is the intrinsic driver for the 100% selectivity to the target *p*-CAN product relative to Ni-Al. The increase of activity might be due to Cu-Ni surface interaction and electron transfer from Cu to Ni [176].

3. Synthesis of Aromatic Amines under Solvent-Free Conditions or in Green Solvents

Organic solvents are usually necessary for the liquid phase hydrogenation and the gas phase hydrogenation of solid substrates. The use of traditional organic solvents is a major source of waste in chemical industries and many of them are associated with serious health hazards and/or environmental problems [179-181]. The hydrogenation of aromatic nitro compounds has been examined in some green systems such as reactions under solvent-free conditions, in ionic liquids (ILs), $scCO_2$, CXLs, and water.

3.1. Solvent-Free Hydrogenation

Solvent-free reactions have several advantages over reactions in organic solvents. The concentration of substrate is higher and the size of reactor is smaller. The pollution from organic solvents is also eliminated. The separation and purification of products is easier and energy consumed is less than that required in organic solvents.

Li et al. prepared a heterogeneous catalytic assembly of Pd nanoparticles in SiO_2-supported ionic liquid brushes. The hydrogenation of NB to AN could proceed over the catalyst under solvent-free conditions at room temperature and at normal pressure, giving an almost 100% yield. The catalyst could be recycled 15 times without apparent loss of activity [182].

The solvent-free hydrogenation of NB and *o*-CNB to corresponding aromatic amines was performed over Pt nanoparticles supported on ceria nanowires (Pt-CeO_2) at 333 K. The TOF of NB conversion was calculated to be 20700 h^{-1} at 2 MPa H_2 and that of *o*-CNB hydrogenation to be 8358 h^{-1} at 3 MPa H_2. In the hydrogenation of *o*-CNB, the selectivity to *o*-CAN was 99.2% at a 99% conversion [183]. Ultrafine Pt particles (≤3.5 nm) deposited on multi-walled carbon nanotubes (Pt/MWNTs) were also prepared for the solvent-free hydrogenation of NB. At 333 K and at 2 MPa H_2, the activity of Pt/MWNTs (TOF=52000 h^{-1}) was much higher than those of the above Pt-CeO_2 (TOF=20700 h^{-1}) and the commercial Pt/C catalysts (TOF=13500~22300 h^{-1}). The selectivity to AN was >99% at complete conversion. NSB and AOB were the only two intermediates with the yields <3% during the process of NB conversion [42]. Thus, the serious accumulation of harmful intermediates was suppressed over Pt/MWNTs.

A Pt/γ-Fe_2O_3 catalyst with an average size of Pt particles being 2.3 nm was used for the solvent-free hydrogenation of *o*-CNB to *o*-CAN. The selectivity to *o*-CAN was >99.6% at nearly complete conversion of *o*-CNB at 313–333 K and at 0.6–2.0 MPa H_2. However, the support was partially reduced to Fe_3O_4 and Fe(0), and the morphology changed during the reaction. A slight leaching of Fe into the water phase was determined to be about 1% loss from the γ-Fe_2O_3 support [46]. A well dispersed Pt catalyst, with an average diameter of Pt clusters being 1.4±0.3 nm, was prepared by adsorption of Pt colloidal particles on a Fe_3O_4 support (Pt/Fe_3O_4) and used for the solvent-free hydrogenation of CNBs and *o*-BNB. Its activity for the hydrogenation of *o*-CNB was about three times as high as that over Pt/γ-Fe_2O_3 or commercial Pt/C. The selectivity to HANs was >99% at complete conversion of substrates and the yield of aniline was <1%. Prolonging the reaction time after complete conversion of the substrate did not decrease the selectivity to *o*-CAN, indicating that the hydrodechlorination of *o*-CAN was perfectly suppressed. The catalyst was stable enough to be reused four times without apparent loss of activity and selectivity. A slight metal (Pt and Fe) leaching was detected in the aqueous phase of the products possibly due to the corrosion of Fe_3O_4 by a small amount of HCl formed *in situ* [47]. The solvent-free hydrogenation of *o*-CNB to *o*-CAN was also examined over Pt-Ru nanoparticles supported on Fe_3O_4-modified carbon (Pt-Ru/Fe_3O_4/C). The selectivity to *o*-CAN was >99.4% at 100% conversion of *o*-CNB and the yield of dechlorinated products was <0.1%, indicating that the dechlorination was suppressed successfully. The catalyst was stable enough to be recycled 19 times without obvious loss of activity and selectivity. Pt or Ru was not detected in the filtrate. Aggregation of Pt and Ru nanoparticles was observed and might be responsible for the deactivation of Pt-Ru/Fe_3O_4/C after 21 cycles [28]. The above Pt catalysts containing Fe_3O_4 could be easily recovered by an applied magnetic field after the reaction.

The hydrogenation of solid aromatic nitro compounds was also reported to proceed under solvent-free conditions using Pd nanoparticles entrapped in aluminum oxy-hydroxide (Pd/AlO(OH)). The catalyst did not lose it activity significantly after three cycles. The efficiencies of solvent-free hydrogenation of solid substrates by Pd/AlO(OH) were closely dependent on the melting point of reactants. The solvent-free hydrogenation of substrates with lower melting point proceeded faster than those with higher melting point under the same conditions [86].

Solvent-free reactions could be the cleanest and most efficient process if suitable catalysts were available and the industrial operation can be controlled easily. However, the hydrogenation of aromatic nitro compounds is highly exothermic. For example, the heat of hydrogenation of NB to AN is about 493 kJ mol^{-1} [184]. Thus, reaction temperature will be difficult to be controlled during the solvent-free hydrogenation of aromatic nitro compounds in industrial scale.

3.2. Hydrogenation in Ionic Liquids

ILs are fused salts that are composed entirely of cations and anions and melt around 373 K or below [185]. ILs have an extremely low vapor pressure, a wide liquid range, tunable polarity, and potentially easy recyclability [186]. The hydrogenation of halogen substituted nitrobenzene has been examined in ILs that serve as green solvent to replace methanol. Both of the reaction rate and the selectivity to HAN are largely improved with comparing to the results in methanol.

The hydrogenation of CNBs (*o*-, *m*-, and *p*-CNB), *m*-BNB, and 3,4-dichloronitrobenzene in ILs was compared with that in methanol over Raney Ni, commercial Pt/C and Pd/C catalysts, respectively [187]. The applied ILs were 1-alkyl-3-methylimidazolium based tetrafluoroborates ([RMIM][BF_4])

and hexafluorophosphates ([RMIM][PF_6]). Although the reaction rate in [RMIM][BF_4] was lower than that in methanol, the selectivity to HANs was significantly improved because the dehalogenation was markedly suppressed in [RMIM][BF_4]. Both the anions and cations of ILs had a strong impact on the catalytic selectivity and the hydrogenation properties could be tuned by the proper choice of ILs. For example, the debromination could be completely inhibited in [BMIM][PF_6] and [HMIM][PF_6] during the hydrogenation of *m*-BNB over Pd/C that generally promotes extensive dehalogenation in organic solvents. The heterogeneous metal catalyst-[BMIM][BF_4] system was easily separable from the products and the remained catalyst system could be reused at least five times without apparent decrease of activity and selectivity.

The homogeneous hydrogenation of *o*-CNB to *o*-CAN in [BMIM][BF_4] was also examined with an air-stable cationic donor stabilized phosphenium Pt(II) complex as catalyst [188]. The yield of AN byproduct was <0.5% at complete conversion of CNB, indicating that the dehalogenation reaction was suppressed successfully. The catalyst could be reused seven times with a slight decrease of activity and selectivity.

Xiao et al. prepared Pt nanoclusters (1.7 nm) stabilized by an ionic liquid-like copolymer poly[(*N*-vinyl-2-pyrrolidone)-*co*-(1-butyl-3-vinylimidazolium chloride)], which exhibits good solubility in [BMIM][BF_4] ionic liquid. The Pt catalyst exhibited excellent activity and selectivity (≥97.5%) for the hydrogenation of *o*-CNB to *o*-CAN in [BMIM][BF_4] and was recycled with the activity and selectivity preserved [186]. Both the dehalogenation and the accumulation of harmful intermediates were successfully suppressed. Intermolecular interactions between [BMIM][BF_4] and *o*-CNB weakened the N=O bond of nitro group, but not the C-Cl bonds, which might be responsible for the enhanced selectivity to *o*-CAN.

The use of ILs as solvents for the hydrogenation of aromatic nitro compounds is advantageous not only for product separation and catalyst recycling, but also for selectivity improvement. However, an idea should be kept in mind that the synthesis of ILs is complex with using many organic and inorganic substances [189].

3.3. Hydrogenation in Supercritical CO_2

ScCO_2 (T_c=304.25 K, P_c=7.38 MPa) is an attractive alternative to conventional organic solvents, due to its nonflammability, relative inertness, complete miscibility with gases, and easy separation from liquid/solid products after reactions [190-195]. The hydrogenation of several aromatic nitro compounds has been performed in $scCO_2$, and $scCO_2$ showed interesting effects on the reaction rate and product selectivity. For example, in $scCO_2$ the hydrogenation of CNB to CAN over Pt/C and Ni/TiO_2 was largely promoted, but the same reaction over Au/TiO_2 was retarded due to serious deactivation of the Au/TiO_2 [8, 20, 196]. This subsection will review the significance of $scCO_2$ for the hydrogenation of aromatic nitro compounds and summarize recent understanding on the important roles of compressed CO_2 including interaction with metal catalysts and reacting species.

The hydrogenation of aromatic nitro compounds in $scCO_2$ was investigated with several supported transition metal catalysts. Zhao et al. studied the hydrogenation of NB at 308 K in $scCO_2$ with noble metal catalysts including Pd, Pt, Ru, and Rh supported on C, SiO_2, and Al_2O_3 [17]. The catalytic activity followed the order of Pt/C > Pd/C > Ru/C, Rh/C. The supports showed an influence on the activity of both Pd and Pt catalysts in $scCO_2$, and C was more effective than SiO_2 and Al_2O_3. The selectivity to AN was higher in $scCO_2$ than that in ethanol for all the catalysts used. The presence of compressed CO_2 was helpful to retard the formation of harmful by-products such as NSB, AOB, and AB over all the Pt and Pd catalysts (supported on C, SiO_2, and Al_2O_3); the by-products formed over Pd catalysts was less than that over Pt catalysts at conversion levels <100% [17]. Chatterjee et al. studied the hydrogenation of NB in $scCO_2$ at 323 K over MCM-41 supported Pd, Pt, Rh, Ru, Au, Ir, and Ni catalysts, respectively. The initial activity of the catalysts followed the order of Pd > Pt > Au ≈ Ir ≈ Rh ≈ Ru ≈ Ni. The selectivity to AN over Pd/MCM-41 was 100% independent of conversion. The apparent activation energy for the hydrogenation of NB in $scCO_2$ was determined to be 15 $kcal\cdot mol^{-1}$ over Pd/MCM-41 at 308–333 K. These authors proposed that the reaction follows the pathway of NB→PHA→AN, and the hydrogenation of NB to PHA is the rate-determining step [23].

The selective hydrogenation of NB was investigated with Ni/γ-Al_2O_3 using different media of compressed CO_2, ethanol, and *n*-hexane at 308 and 323 K. In ethanol, several undesired toxic intermediates including NSB, PHA, AOB, and AB were formed during the course of reaction. The total rate of NB hydrogenation with Ni/γ-Al_2O_3 in compressed CO_2 at 12 MPa was larger than that in organic solvents under similar reaction conditions; the selectivity to AN was almost 100% over the whole conversion range up to 100%. Thus, the reduction of intermediates can be accomplished along with an increase in the reaction rate in the presence of compressed CO_2 [7].

The CO_2 pressure has significant effects on the total conversion of NB over Pt, Pd, and Ni catalysts. Generally, the conversion increases first and then decreases on further introduction of CO_2; the conversion is maximal at CO_2 pressures corresponding to the transition of the gas–liquid–solid system to the gas–solid one [7, 17, 19, 23]. The decrease in conversion with further introduction of CO_2 may be due to a dilution effect. Thus, the phase behavior is an important factor in determining the rate of the hydrogenation under compressed CO_2. The selectivity to AN changed slightly with CO_2 pressure for Pd/C and Pt/C [17, 19]; in contrast, the perfect selectivity (~100%) to AN remained unchanged for Ni/Al_2O_3 and Pd/MCM-41 even when the reaction mixture changed from the gas–liquid–solid system to the gas–solid one [7, 23].

Zhao et al. also studied the influence of metal particle size on the hydrogenation of NB in scCO_2. The catalysts they used were commercially available Pd/C, Pd/Al_2O_3, Pt/C, and Pt/Al_2O_3 (Wako) and silica-supported Pd and Pt prepared by ion-exchange method. The dispersion of metal was changed from ca. 0.03 to 0.6. The hydrogenation over all Pd and a few selected Pt catalysts is structure sensitive. An exception was the reaction on Pt/C with Pt dispersion ranging from 0.04 to 0.13; the hydrogenation was structure insensitive in scCO_2 [17, 19].

The presence of compressed CO_2 is helpful to improve the selectivity to the desired HANs during the hydrogenation of HNBs with several supported metal catalysts. Ichikawa et al. examined the hydrogenation of *o*-CNB over Pt/Al_2O_3, Pt/SiO_2, and charcoal supported Pt, Pd, Ir, Rh, and Ru catalysts. The catalytic activity followed the order of Pt/C > Pd/C > Ir/C > Rh/C, Ru/C, and the performance of Pt/C was better than Pt/Al_2O_3 and Pt/SiO_2. These authors compared the hydrogenation of *o*-CNB over 1 wt% Pt/C in the presence and absence of 10 MPa CO_2 at 313 and 363 K. The use of scCO_2 enhanced the reaction rate and strongly suppressed the dechlorination, resulting in a marked improvement in the selectivity to *o*-CAN compared with the neat reaction [20]. The undesired dechlorination could be almost completely suppressed for Pt/C, Ir/C, Pt/Al_2O_3, and Pt/SiO_2 in scCO_2, but not for Pd/C [20, 21]. The reaction system is also effective for the selective hydrogenation of other aromatic nitro compounds such as *m*-, *p*-CNB, and 2,5-dichloronitrobenzene [20]. The hydrogenation of CNBs to CANs was investigated over Ni/TiO_2 at 308 K in scCO_2, ethanol, and *n*-hexane. The reaction rate followed the order of scCO_2 > *n*-hexane > ethanol. In scCO_2, the selectivity to CANs and to AN over Ni/TiO_2 were 97–99.5% and <1%, respectively, in the conversion range of 9–100%. The high chemoselectivity to CANs cannot be achieved over Ni/TiO_2 in ethanol and *n*-hexane [8].

Zhao et al. examined the hydrogenation of several substituted nitro compounds such as *o*-, *m*-, *p*-nitroanisole, *o*-, *m*-, *p*-nitrotoluene, 1,3-dinitrobenzene and 2,4-dinitrotoluene in $scCO_2$ and in ethanol with 5 wt% Pt/C at 323 K. Under the same concentrations of substrate and catalyst, the conversion in $scCO_2$ was comparable with that in ethanol; however, the selectivity to aromatic amines was larger in $scCO_2$ than that in ethanol [18]. The above-mentioned Pd/MCM-41 catalyst is also effective for the selective hydrogenation of nitrotoluene, nitroanisole, and nitrobenzoic acid isomers in $scCO_2$, giving ~100% selectivity to aromatic amines at any conversion level. The conversion of isomers followed the order of $p > m > o$. However, the catalyst is not good for the hydrogenation of *m*-CNB to *m*-CAN due to a relatively high level of dechlorination [23].

The above results indicate that $scCO_2$ is a suitable replacement for organic solvents in the hydrogenation of some aromatic nitro compounds. Compressed CO_2 may enhance the solubility of reagent gases, increase the rate of transport, lower the melting point of substrates, generate *in situ* acidic species, and provide some environmental and safe benefit. Herein, discussion on the role of compressed CO_2 will be given to the interactions of CO_2 with the catalyst surface and reacting species.

Baiker and co-workers showed that CO formation via the reverse water-gas shift reaction from $scCO_2$ and H_2 ($H_2 + CO_2 \rightarrow CO + H_2O$) was a general feature of the commonly used Pt-group metal catalysts (Pt, Pd, Ru, and Rh) [197, 198]. Ichikawa et al. reported FTIR spectra of CO generated and adsorbed on the Pt surface during the reaction of CO_2 with H_2 under atmospheric conditions. It was suggested that the *in situ* formed CO may be adsorbed preferentially on the kink or step sites of the Pt surface that are active for the dechlorination. Thus, CO_2 may be a source of surface modifier (CO) for Pt/C in the hydrogenation of CNBs [20]. The formation and adsorption of CO were also observed for commercial Rh/C and Rh/Al_2O_3 catalysts at high CO_2 pressures in the presence of H_2. This was considered to be one of important factors retarding the rate of hydrogenation of phenol in the presence of CO_2 [199]. Hao et al. found that the Au/TiO_2 catalyst deactivated during the hydrogenation of *o*-CNB in $scCO_2$. Carbonate-like species are formed on gold and may cover some of active sites. The original Au^0 species was partially oxidized by CO_2 even in the presence of H_2 [196].

The selectivity to the desired aromatic amines can be improved over various metal catalysts such as Ni, Pt, Pd, and Ir when the hydrogenation of aromatic nitro compounds is performed in/under compressed CO_2 [7, 8, 17, 19-21]. Meng et al. studied the molecular interactions of CO_2 with reactant and reaction intermediates, e.g. NB, NSB, and PHA, in the hydrogenation of NB. Their results show that compressed CO_2 strongly interacts with these compounds and modifies the reactivity of each species. For NB, the bands of the symmetric and asymmetric stretching vibrations of the nitro group blue shifted with CO_2 pressure up to ca. 6 MPa. At 6–20 MPa CO_2, these peaks slightly red shifted but were located at larger wavenumbers than those in the ambient gas state and in the liquid state. The authors proposed two interaction modes between CO_2 molecules and the nitro group (Scheme 2). The mode A is responsible for the large blue shift and the mode B causes the slight red shift at 6–20 MPa CO_2. The IR spectra of NSB showed that it dominantly interacted by its oxygen atom to the carbon of CO_2 (mode A in Scheme 3) and this type of interaction weakened the N=O bond of NSB with the increase in CO_2 pressure. Interactions also occurred between PHA and compressed CO_2 molecules. In a word, the reactivity of NB is decreased but that of NSB is increased and the transformation of PHA to AN is likely promoted. The interactions of CO_2 with the reacting species are an important factor responsible for the improved selectivity to AN in/under compressed CO_2. These authors proposed that the hydrogenation of NB in compressed CO_2 over Ni/γ-Al_2O_3 mainly occurs through the consecutive route, i.e., NB→NSB→PHA→AN. The transformation of NB to NSB might be the rate-determining step [7]. The molecular interactions of CO_2 with reactants and intermediates are also important for the selective hydrogenation of CNBs to CANs over Ni/TiO_2 in compressed CO_2 [8].

Although scCO_2 is a green and effective medium for the hydrogenation of several aromatic nitro compounds, its application is sometimes limited by the low solubility of some substrates in scCO_2. High CO_2 pressures (>10 MPa) are commonly required to achieve the maximum reaction rate.

(A) (B)

Scheme 2. Possible interaction modes of compressed CO_2 with the nitro group.

Scheme 3. Possible interaction modes of compressed CO_2 with the nitroso group.

3.4. Hydrogenation in CO_2-Expanded Liquids

CXLs are a new class of interesting reaction media. A CXL can occur when an organic solvent or substrate is pressurized by CO_2 and a large quantity of CO_2 is dissolved into the liquid phase. Compared to reactions in $scCO_2$, the reaction in CXLs is not limited by the solubility of reacting species in $scCO_2$ and can process larger amounts of substrates. The catalysts active in organic liquid and aqueous phases can be used without modifications to increase their solubility in $scCO_2$. The required CO_2 pressure in CXLs is lower than that in $scCO_2$ [200, 201].

The hydrogenation of NB was performed over Pd/C in CO_2-expanded cyclohexane at 323 K. At 4–8 MPa CO_2, the conversion of NB (>90%) in the CO_2-expanded phase was larger than that in the neat cyclohexane (~70%); the selectivity to AN was a little improved in CO_2-expanded cyclohexane (>95%) but slightly lower than that in $scCO_2$ (~100%) [26]. The hydrogenation of *p*-CNB with a Ni–B nanocatalyst was examined in CO_2-expanded methanol at 343–363 K over a total pressure range of 3.5–7.1 MPa. The conversion of *p*-CNB in CO_2-expanded methanol was higher than that in unexpanded methanol [25].

Meng et al. conducted the hydrogenation of NB over Ni/Al_2O_3 in CO_2-expanded ethanol, *n*-hexane, and water at 323 K. The pressurization with CO_2 had a drastically negative impact and a slightly positive effect on the conversion of NB in ethanol and in *n*-hexane, respectively. When the H_2O–NB mixture was compressed by CO_2 at a low pressure such as 0.8 MPa, the conversion was significantly enhanced and comparable to that obtained in the supercritical 12 MPa CO_2–NB mixture.

The selectivity to AN was almost 100% in the whole range of conversion for the reactions in CO_2–*n*-hexane and CO_2–H_2O systems. The reaction system including H_2O and low-pressure CO_2 was also effective for the selective hydrogenation of CNBs to CANs over Ni/TiO_2 at 308 K. These authors proposed that the hydrogenation of nitro group should be the rate-determining step. Thus, the hydrogenation of some water-insoluble aromatic nitro compounds could be performed selectively in a CO_2–H_2O system at CO_2 pressures lower than its critical pressure [9].

Fujita et al. studied the effects of pressurized CO_2 on the selective hydrogenation of nitrostyrene to aminostyrene over Pt/TiO_2 at 323 K and at 4 MPa H_2. The conversion of nitrostyrene and the product distribution were both affected by CO_2 pressurization. In the CO_2-expanded nitrostyrene phase, the over-all reaction rate and the selectivity to aminostyrene increased with CO_2 pressure up to 10 MPa. At higher CO_2 pressures where all nitrostyrene dissolved in scCO_2 (scCO_2–homogeneous phase), both of them decreased with increasing the CO_2 pressure. They found the presence of pressurized CO_2 suppresses the hydrogenation of aminostyrene to ethylaniline under CXL conditions. These authors further investigated molecular interactions of pressurized CO_2 with nitrostyrene by FTIR. Their results showed that pressurized CO_2 interacted only with the nitro group of nitrostyrene and lowered its reactivity; the interaction was stronger in scCO_2–homogeneous phase than in CXL phase.

In other words, the reactivity of the vinyl group is unchanged but the reactivity of the nitro group is much lowered by pressurized CO_2, resulting in the lower selectivity to aminostyrene in the scCO_2–homogeneous phase. Thus, the CXL conditions are better than scCO_2 homogeneous conditions for the selective hydrogenation of nitrostyrene to aminostyrene. FTIR measurements of adsorbed CO over the catalysts suggested that the presence of less-coordinated Pt atoms on edge, corner and kink sites was significant for the selective hydrogenation of nitrostyrene to aminostyrene [22].

The hydrogenation of nitrostyrene was also studied over Pt/TiO_2 in CO_2-expanded ethanol and toluene phases at 323 K. The conversion simply decreased with CO_2 pressure in the both expanded solvents. In ethanol, the product distribution at 0.1 MPa CO_2 was similar to that at 10 MPa and aminostyrene was the main product with ca. 60% selectivity over the whole range of conversion. In toluene, the selectivity to aminostyrene at 0.1 MPa CO_2 was ca. 60% at conversion higher than 60%; however, this value at 10 MPa CO_2 was <15% over the whole range of conversion and the main product was ethylnitrobenzene.

FTIR measurements showed that the extent of blue shift of $\nu_{as}(NO_2)$ in CO_2-expanded toluene (ca. 4 cm^{-1}) was larger than that in CO_2-expanded ethanol (ca. 2 cm^{-1}) at 0.1–8 MPa; the dissolved CO_2 also affected the absorption of the OH bond of the ethanol solvent. Combined with results from molecular dynamics simulation, these authors proposed that the local composition around a nitrostyrene molecule changed little for the CO_2-dissolved ethanol phase even at high CO_2 pressures but did change for the CO_2-dissolved toluene phase.

In the latter phase, more CO_2 molecules occupy the nearest neighbor sites around a nitrostyrene substrate at higher pressure. This induces the significant interactions of its nitro group with CO_2 molecules, resulting in a decrease in the reactivity of the nitro group. Such an interaction is negligible in the case of ethanol because the interactions of the nitro group with ethanol through its hydroxyl group are more significant. Thus, the pressurization with CO_2 varies the product selectivity in the nonpolar toluene but not in the polar ethanol [27].

3.5. Hydrogenation in Water

Water is the most abundant, natural and non-toxic solvent with high heat capacity and unique redox stability on earth. Water has a network of hydrogen bonds and has the ability to form hydrogen bonds with some substrates, which can influence the reactivity of reacting species. When the substrate of a reaction is practically insoluble in water, the reaction still may proceed, in some cases, with higher synthetic efficiency due to the "on water" effects. In large industrial processes, isolation of water-insoluble organic products can be performed by simple phase-separation [181, 202, 203]. The hydrogenation of aromatic nitro compounds is highly exothermic, and most of aromatic nitro reactants and their hydrogenation products, aniline species, have low solubility in water.

Thus, besides the merits of cost and environmental benign nature, the use of water as a medium for this reaction has other advantages including facile control of reaction temperature, easy separation of products, and possible promoting effects of water. Present results show that water does promote the hydrogenation of aromatic nitro compounds. The promotion effects of H_2O have been mainly attributed to the interaction of H_2O with reacting species such as NB and PHA.

The hydrogenation of aromatic nitro compounds with homogeneous catalysts has been examined in neat H_2O or in aqueous/organic biphasic systems. The hydrogenation of NB with Ru(II) phenanthroline complex as a catalyst was performed in H_2O, methanol, isopropanol, and *n*-butanol, respectively. Both the conversion of NB and the selectivity to AN in H_2O were higher than those in the above alcohols. The hydrogenation of other aromatic nitro compounds such as *p*-nitrobenzaldehyde, *p*-nitrobenzoic acid, and *p*-CNB could be also carried out with the catalyst in H_2O, giving corresponding aromatic amines with high chemoselectivity (>99%) [204]. The hydrogenation of *p*-CNB with a water-soluble Pd-lysine complex catalyst was examined in H_2O at 298–348 K and at 0.1–0.4 MPa H_2. The selectivity to *p*-CAN was ≥98% at 0.1 MPa H_2 and at 298–323 K [205]. Huang et al. prepared water-soluble amphiphilic tannin-stabilized Pd(0) nanoparticles that are active for the biphasic hydrogenation of hydrobic substrates, e.g. NB and nitrotoluene. The recovery of the catalyst is easy because tannin polymer constrained Pd particles in the aqueous phase. The amphiphilic tannin polymer can also form hydrophobic bonds with the hydrophobic organic phase, thus decreasing the interfacial resistance [206].

Maity et al. prepared Pt nanoparticles (2–8 nm) supported on a water-soluble polymer poly(diallyldimethylammonium chloride) and examined its catalytic performance for the hydrogenation of NB, *o*- and *m*-CNB, and *p*-NAN in H_2O and organic solvents at 300 K and at 5.0 MPa H_2. The hydrogenation of water-insoluble NB in H_2O was faster than that in methanol. For the hydrogenation of *o*- and *m*-CNB, and *p*-NAN solids that are practically insoluble in H_2O, turnover numbers in H_2O were much higher than those in a 1:1 mixture of H_2O and toluene. The hydrogenation of *o*-CNB in H_2O yielded only *o*-CAN; no AN and any other byproducts were detected [207]. Sreedhar et al. prepared Pt nanoparticles (1–4 nm) in H_2O using water-soluble gum acacia (GA) as reducing and stabilizing agent. The colloidal Ga-Pt nanoparticles catalyzed the hydrogenation of several aromatic nitro compounds (e.g., NB, nitrotoluene, nitroanisole, nitrophenol, HNBs, and 4-nitrobenzaldehyde) to corresponding aromatic amines in good to excellent isolated yields (68–95%) in H_2O at room temperature under 0.1 MPa H_2. Dehalogenation reaction was not observed for the hydrogenation of HNBs. The GA–Pt colloidal solution was reused for several times with moderate loss in catalytic activity (91% yield in the first cycle and 84% in the fifth cycle) [208].

Pd nanoparticles (3–6 nm) dispersed on porous poly(vinylidene fluoride) (PVDF) were used to heterogeneously catalyze the hydrogenation of NB to AN in H_2O at room temperature and at normal pressure. The PVDF polymer is water insoluble but water dispersible, which affords both heterogeneous nature and good dispersion of the catalyst in H_2O.

The catalyst could be separated easily by filtration and reused three times without aggregation of Pd particles [209]. Pd nanoparticles in SiO_2-supported ionic liquid brushes were synthesized for the hydrogenation of aromatic nitro compounds in neat H_2O at 303 K and at 0.1 MPa H_2.

The catalyst worked well with both liquid and solid substrates. When the reduction products are also water-insoluble solids at reaction temperature, the hydrogenation reactions proceeded in a lower rate. However, the catalytic system is not suitable for the hydrogenation of HNBs because serious dehalogenation reaction occurs over it [182].

The hydrogenation of NB, *p*-nitrotoluene, *p*-CNB, and *p*- and *o*-nitrophenol was carried out in H_2O with Ni nanoparticles stabilized by a hydrophilic ionic liquid 1-(3-aminopropyl)-2,3-dimethylimidazolium acetate ([AMMIM][AcO]). At 343–353 K and at 1 MPa H_2, the selectivity to corresponding aromatic amines was >99% at 100% conversion.

The catalyst could be reused five times without apparent loss of activity and selectivity. However, serious accumulation of intermediates, e.g., NSB and AB, occurred during the conversion of NB [210].

The hydrogenation of NB over Ni/TiO_2 was performed in H_2O at 353 K and at 6.5 MPa H_2. The selectivity to AN was >98.5% at any conversion level and no undesired intermediates accumulated during the reaction. However, the catalyst rapidly deactivated due to the formation of Ni hydroxide from metallic Ni reacting with H_2O. The activity and stability of Ni/TiO_2 was largely improved by coating a hydrophobic carbon layer on the surface of Ni/TiO_2. Ni hydroxide was not detected on the used and H_2O treated carbon-coated Ni/TiO_2. The carbon-coated Ni/TiO_2 could be recycled three times with a slight decrease in reaction rate [159].

The promoting role of H_2O in the hydrogenation of aromatic nitro compounds is a question of common interest. Jorge et al. performed a systematic molecular dynamics study of the H_2O/NB interface and proposed that H_2O molecules at the interface establish hydrogen bonds via their hydrogen atoms with NB oxygen [211].

The hydrogen bonds may weaken the N–O bond of NB. Hydrogenation run of NB, NSB, and PHA was respectively conducted over Ni/Al_2O_3 in H_2O at 323 K and at 6 MPa H_2. The hydrogenation rate followed the order PHA > NB > NSB. The molecular interactions of H_2O with a hydrophilic intermediate of PHA were examined by FTIR and the results show that the N–O bond of PHA is weakened through interactions with H_2O, possibly *via* OH···O and OH···N bonding. Thus, water can promote the reaction step of PHA to AN [9].

4. Other Important Progresses in the Reduction of Aromatic Nitro Compounds

4.1. Coupling of the Hydrogen Production with the Hydrogenation

Hydrazine hydrate, formates and $NaBH_4$ as hydrogen source were extensively used in the reduction of aromatic nitro compounds, and these reactions as probe reactions were usually used to check up the catalytic performance of novel catalysts [212-216], although the atom economy of the reactions with these hydrogen sources was lower than that of the reactions with hydrogen gas. It will not be involved, but the coupling of the hydrogen production with hydrogen sources like *i*-propanol with the hydrogenation, and coupling of the dehydrogenation and hydrogenation reactions will be related herein.

Alcohol especially *i*-propanol as hydrogen source under basic conditions (NaOH, KOH etc.) was used in the reduction of aromatic nitro compounds, however, condensation route is favored under these conditions [3], resulting in low selectivity to aromatic amines. Li et al. used ethanol or methanol/Raney Ni, methanol/Pd-Fe/Al_2O_3 or Pd-Ba/Al_2O_3 to catalyze the reduction of NB, but the selectivity to AN was low (<22%) in the absence of base because the formed AN condensed with aldehyde that was transformed from alcohol [217-219]. However, in the presence of water, aqueous-phase reforming of alcohol produced CO_2 and H_2 but not the aldehyde, the selectivity to AN increased to >94% at conversion >30% [219-221]. In addition, pinacol was reported as a new green reducing agent for the reduction of aromatic nitro compounds with dichlorodioxomolybdenum (VI) complexes as catalysts, a wide range of substrates could be reduced to corresponding aromatic amines at yield >70% [222]. Moreover, dehydrogenation of 1,4-butanediol to γ-butyrolactone or ethylbenzene to styrene could be coupled with the hydrogenation of NB to produce AN [223, 224].

The H_2 produced from CO-H_2O water–gas shift (WGS) reaction can be used in the hydrogenation. In early literature, $Rh_4(CO)_{12}$-9,10-diamino-phenanthrene [225], cis-[$Rh(CO)_2$(2-picoline)$_2$](PF_6) [226], and sulfonated 2,2'-bis diphenylphosphinomethylene binaphthyl/$PdCl_2$ [227] catalysts had been reported for the hydrogenation of aromatic nitro compounds, unfortunately, other functional groups, like Cl, C=O, C=C, were reduced simultaneously, and the yield of corresponding aromatic amines was low [225,

227]. Supported Au catalysts have been found to be very active for WGS reaction [228, 229]. Thus they may show high activity and even selectivity for the hydrogenation of aromatic nitro compounds by combining with the low temperature WGS reaction. Deng et al. reported that ferric hydroxide supported nanocluster gold could catalyze the hydrogenation of aromatic nitro compounds (Cl, Br, OMe, C=O substituted NB) to corresponding aromatic amines (S >98%, C >95%, TOF >312 h^{-1}) in the presence of CO and H_2O at 373-393 K [230]. Cao et al. reported that Au/TiO_2 could reduce aromatic nitro compounds (F, Cl, C=O, C=C, NO_2 etc. substituted NB) to produce corresponding aromatic amines (S >91%, C >96%, TOF >33 h^{-1}) with CO and H_2O even at room temperature [231]. Thereafter, Kaneda et al. reported that CO/H_2O as a reducing reagent could be used to reduce *m*-nitrostyrene to *m*-aminostyrene (S >99% at C >99%) over the hydrotalcite-supported silver nanoparticles catalyst (Ag/HT). The yield over Ag/HT was higher than that over HT supported Au, Pd, Rh and Pt, and there was no accumulation of azobenzene compounds [136]. In addition, it was reported that selenium powder [232] and Ru bis(arylimino)acenaphthene complexes [233] could catalyze the reduction of aromatic nitro compounds to aromatic amines with CO and H_2O reducing reagent.

4.2. Photocatalytic Reduction of Aromatic Nitro Compounds

Photocatalysis has been paid more attention in the organic synthesis. When semiconductor catalyst is irradiated by UV light, charge separation occurs and thus-formed electrons (e^-) in the conduction band and holes (h^+) in the valence band cause reduction and oxidation, respectively. The reduction reaction can be carried out in the presence of a large excess of a sacrificial electron donor (SED), which plays an important role in scavenging valence band holes in order to limit the charge recombination degree and then liberating more photogenerated electrons [234]. The formed photogenerated electrons can be used for reduction of aromatic nitro compounds. Many solvents, such as alcohols, cyclohexene, organic acids, and their mixtures have been reported as useful SED in improving the photoredox cycle [234-242]. The catalysts such as N-doped TiO_2 [242, 243], amino acid modified TiO_2 nanoparticles [244, 245], Ag/TiO_2 [241], and nanocrystalline $PbBi_2Nb_2O_9$ [246] were also reported for the photocatalytic reduction of aromatic nitro compounds. Alcohols were oxidized to aldehydes and aromatic nitro compounds were reduced to aromatic amines, but the condensation of aldehydes and aromatic amines resulted in low selectivity to aromatic amines [234, 240]. Dioxygen (O_2) was sacrificial electron acceptor to consume e^-, so removal of O_2 from the system was an absolutely necessary process to improve the reduction efficiency by removing this competitive electron scavenger. However, Kominami et al. [247-249] reported high yield (>95%) of aromatic amines could be obtained from the hydrogenation of aromatic nitro compounds over an effective photocatalytic system with TiO_2 combined with HCOOH or $(COOH)_2$ in water, acetonitrile, or the mixture of water and acetonitrile under air or argon. No dehalogenation reaction occurred for the HNBs hydrogenation, and the reducible substituent groups such as olefins and ketones remained intact

4.3. One Pot Synthesis of Imines and Secondary Aromatic Amines

Imines, secondary amines and their derivatives are versatile intermediates in organic synthesis. Imines derivatives can be obtained by condensation of aromatic amines and aldehydes, and secondary amines can be synthesized by the hydrogenation of imines [250-252]. However, it is interesting to obtain imines or secondary amines from aromatic nitro compounds by one pot reaction, and some research had been reported in literature. Ma et al. reported one pot synthesis of imines by coupling the hydrogenation of aromatic nitro compounds to aromatic amines over Ni/SiO_2 with the condensation of aromatic amines and benzaldehydes [253]. Imines derivatives can be also obtained by one-pot reduction of aromatic nitro compounds using aldehydes in the presence of carbon monoxide and water (CO-H_2O water–gas shift reaction) [254, 255] or in the presence of methanol and water (aqueous-phase reforming of alcohol) [256]. One pot synthesis of secondary amines from aromatic nitro compounds and benzaldehydes could be catalyzed by Pt [250, 257], Ru [258], Pd [259], and Au [251, 260] catalysts. Usually, this reaction goes through three steps: reduction of aromatic nitro compounds to aromatic amines, then condensation of aromatic amines and benzaldehydes to imines, whereafter reduction of imines to secondary amines. However, the reduction rate of imines to secondary amines was much lower than that of other two steps, resulting in the accumulation of imines in the synthesis of secondary amines [250, 251]. At present, Gu and Lu reported an interesting result with ultrathin Pt nanowire catalysts that only trace amounts of *N*-benzylideneaniline (*N*-BIAN), phenylmethanol, and AN were detected with increasing of conversion of NB and yield of *N*-benzylaniline (*N*-BAN). The results of time-dependent analysis, react IR spectroscopy and DFT calculations suggested a new mechanism for the direct reductive C-N coupling. *N*-BAN was directly obtained by intermolecular dehydration over platinum nanowires under a hydrogen atmosphere, instead of intramolecular water elimination of phenyl(phenylamino)methanol and then *N*-BIAN hydrogenation. The excellent performance may be ascribed to the facet (111) that is the predominant plane of bulk Pt crystals [257].

One pot synthesis of secondary amines from aromatic nitro compounds and alcohols could be catalyzed by Ir-Pd [261], Ru [262-264], and Au [265] catalysts. Usually, this reaction goes through four steps: production of hydrogen from the reaction of alcohols to aldhydes, then reduction of aromatic nitro compounds to aromatic amines, thereafter condension of aldhydes and aromatic amines to imines, finally reduction of imines to second amines. In addition, second amines could be obtained from aromatic nitro compounds and nitriles over Pd/C catalyst under 0.1 MPa H_2 at room temperature by one pot reaction [266].

Conclusion

In this decade, the synthesis of aromatic amines from the hydrogenation of aromatic nitro compounds has obtained significant achievements, especially in the development of new catalysts and green reaction processes. For example, the efficient heterogeneous Au-, Ag-based catalysts and supported Ni catalysts as well as the bimetallic catalysts presented higher selectivity compared to the conventional Pt- and Pd-based catalysts, and the aromatic amines such as aminostyrene, aminobenzaldehyde, and haloanilines have been synthesized with high chemoselectivity. The poisonous intermediates such as nitrosobenzene, *N*-phenylhydroxylamine, azoxybenzene, azobenzene, and hydrazobenzene were prohibited efficiently. Moreover, the dehalogenation in the hydrogenation of halonitrobenzenes was retarded and the atom economic efficiency was remarkably enhanced. The fundamental research on the strong metal support interaction has provided strategies for designing efficient catalysts. In addition, the green reaction media such as $scCO_2$, CO_2-H_2O, and CO_2-expanded organic liquids have provided prospective green route for synthesis of aromatic amines, as some unexpected results have been achieved. The dehalogenation and the accumulation of harmful intermediates can be suppressed over Ni/TiO_2 in $scCO_2$ due to the molecular interactions between

CO_2 and substrates; the water-insoluble nitrobenzene and chloronitrobenzene reduction can be promoted in a H_2O–compressed CO_2 system over Ni catalysts at mild conditions. From the standpoint of the industrial utilization, the most prospective synthesis route is hydrogenation of nitro compounds in the green solvents like $scCO_2$, H_2O, and CO_2-expanded liquids; the key problem is to develop stable Ni-based catalysts in the above reaction systems, and the bimetallic catalysts are worth being expected.

Acknowledgments

The authors gratefully acknowledge the financial supports from NSFC 21273222, 21143001 and 21202159, and from Jilin province, China 20100562.

References

[1] A.S. Travis, In: The Chemistry of Anilines; Kosak J. R. Ed.; John Wiley and Sons, (2007) 715.
[2] E.A. Gelder, S.D. Jackson and C.M. Lok, *Chem. Commun.*, (2005) 522.
[3] H.U. Blaser, *Science*, 313 (2006) 312.
[4] A. Corma, P. Concepcion and P. Serna, *Angew. Chem. Int. Ed.*, 46 (2007) 7266.
[5] P. Baumeister, H.U. Blaser and M. Studer, *Catal. Lett.*, 49 (1997) 219.
[6] F. Cardenas-Lizana, S. Gomez-Quero and M.A. Keane, *Appl. Catal. A: Gen.*, 334 (2008) 199.
[7] X.C. Meng, H.Y. Cheng, Y. Akiyama, Y.F. Hao, W.B. Qiao, Y.C. Yu, F.Y. Zhao, S. Fujita and M. Arai, *J. Catal.*, 264 (2009) 1.
[8] X.C. Meng, H.Y. Cheng, S. Fujita, Y.F. Hao, Y.J. Shang, Y.C. Yu, S.X. Cai, F.Y. Zhao and M. Arai, *J. Catal.*, 269 (2010) 131.
[9] X.C. Meng, H.Y. Cheng, S. Fujita, Y.C. Yu, F.Y. Zhao and M. Arai, *Green Chem.*, 13 (2011) 570.
[10] A. Corma and P. Serna, *Science*, 313 (2006) 332.
[11] T. Mitsudome, Y. Mikami, M. Matoba, T. Mizugaki, K. Jitsukawa and K. Kaneda, *Angew. Chem. Int. Ed.*, 51 (2012) 136.
[12] X.D. Wang, M.H. Liang, J.L. Zhang and Y. Wang, *Curr. Org. Chem.*, 11 (2007) 299.

[13] M. Takasaki, Y. Motoyama, K. Higashi, S.H. Yoon, I. Mochida and H. Nagashima, *Org. Lett.*, 10 (2008) 1601.

[14] M.H. Liang, X.D. Wang, H.Q. Liu, H.C. Liu and Y. Wang, *J. Catal.*, 255 (2008) 335.

[15] J.L. Zhang, Y. Wang, H. Ji, Y.G. Wei, N.Z. Wu, B.J. Zuo and Q.L. Wang, *J. Catal.*, 229 (2005) 114.

[16] A. Corma, P. Serna, P. Concepcion and J.J. Calvino, *J. Am. Chem. Soc.*, 130 (2008) 8748.

[17] F.Y. Zhao, R. Zhang, M. Chatterjee, Y. Ikushima and M. Arai, *Adv. Synth. Catal.*, 346 (2004) 661.

[18] F.Y. Zhao, S. Fujita, J.M. Sun, Y. Ikushima and M. Arai, *Catal. Today*, 98 (2004) 523.

[19] F. Zhao, Y. Ikushima and M. Arai, *J. Catal.*, 224 (2004) 479.

[20] S. Ichikawa, M. Tada, Y. Iwasawea and T. Ikariya, *Chem. Commun.*, (2005) 924.

[21] C.Y. Xi, H.Y. Cheng, J.M. Hao, S.X. Cai and F.Y. Zhao, *J. Mol. Catal. A: Chem.*, 282 (2008) 80.

[22] S. Fujita, H. Yoshida, K. Asai, X.C. Meng and M. Arai, *J. Supercrit. Fluids*, 60 (2011) 106.

[23] M. Chatterjee, A. Chatterjee, H. Kawanami, T. Ishizaka, T. Suzuki and A. Suzuki, *Adv. Synth. Catal.*, 354 (2012) 2009.

[24] M. Ohde, H. Ohde and C.M. Wai, *Langmuir*, 21 (2005) 1738.

[25] Y.C. Chen and C.S. Tan, *J. Supercrit. Fluids*, 41 (2007) 272.

[26] J.M. Hao, C.Y. Xi, H.Y. Cheng, R.X. Liu, S.X. Cai, M. Arai and F.Y. Zhao, *Ind. Eng. Chem. Res.*, 47 (2008) 6796.

[27] H. Yoshida, K. Kato, J.Y. Wang, X.C. Meng, S. Narisawa, S. Fujita, Z.J. Wu, F.Y. Zhao and M. Arai, *J. Phys. Chem. C*, 115 (2011) 2257.

[28] Y.A. Wang, Y.X. Yang, Y.W. Li, J.H. Lai and K.P. Sun, *Catal. Commun.*, 19 (2012) 110.

[29] C. Liu, R. Tan, N.Y. Yu and D.H. Yin, *Microporous Mesoporous Mater.*, 131 (2010) 162.

[30] P. Serna, P. Concepcion and A. Corma, *J. Catal.*, 265 (2009) 19.

[31] D.P. He, X.D. Jiao, P. Jiang, J. Wang and B.Q. Xu, *Green Chem.*, 14 (2012) 111.

[32] M.M. Telkar, J.M. Nadgeri, C.V. Rode and R. Chaudhari, *Appl. Catal. A: Gen.*, 295 (2005) 23.

[33] H. Li, H. Lin, Y. Hu, H.X. Li, P. Li and X.G. Zhou, *J. Mater. Chem.*, 21 (2011) 18447.

[34] B. Coq, A. Tijani, R. Dutartre and F. Figueras, *J. Mol. Catal.*, 79 (1993) 253.
[35] X.X. Han, R.X. Zhou, G.H. Lai and X.M. Zheng, *Catal. Today*, 93-5 (2004) 433.
[36] Y.F. Zhao, H.Y. Zhang, C.L. Huang, S. Chen and Z.M. Liu, *J. Colloid Interface Sci.*, 374 (2012) 83.
[37] X.S. Xu, X.Q. Li, H.Z. Cu, Z.B. Huang and X.H. Yan, *Appl. Catal. A: Gen.*, 429 (2012) 17.
[38] X.D. Wang, M.H. Liang, H.Q. Liu and Y. Wang, *J. Mol. Catal. A: Chem.*, 273 (2007) 160.
[39] F. Wang, J. Liu and X. Xu, *Chem. Commun.*, (2008) 2040.
[40] H.W. Ma, K.P. Sun, Y.W. Li and X.L. Xu, *Catal. Commun.*, 10 (2009) 1363.
[41] M.Y. Kim, Y.S. You, H.S. Han and G. Seo, *Catal. Lett.*, 120 (2008) 40.
[42] Z.Y. Sun, Y.F. Zhao, Y. Xie, R.T. Tao, H.Y. Zhang, C.L. Huang and Z.M. Liu, *Green Chem.*, 12 (2010) 1007.
[43] K. Mobus, D. Wolf, H. Benischke, U. Dittmeier, K. Simon, U. Packruhn, R. Jantke, S. Weidlich, C. Weber and B.S. Chen, *Top. Catal.*, 53 (2010) 1126.
[44] Y. Motoyama, Y. Lee, K. Tsuji, S.H. Yoon, I. Mochida and H. Nagashima, *Chemcatchem*, 3 (2011) 1578.
[45] A.J. Kasparian, C. Savarin, A.M. Allgeier and S.D. Walker, *J. Org. Chem.*, 76 (2011) 9841.
[46] M.H. Liu, X.X. Mo, Y.Y. Liu, H.L. Xiao, Y. Zhang, J.Y. Jing, V.L. Colvin and W.W. Yu, *Appl. Catal. A: Gen.*, (2012) doi: 10.1016/j.apcata.2012.07.006.
[47] C. Lian, H.Q. Liu, C. Xiao, W. Yang, K. Zhang, Y. Liu and Y. Wang, *Chem. Commun.*, 48 (2012) 3124.
[48] Y.W. Ma, B. Yue, L.S. Yu, X.Z. Wang, Z. Hu, Y.N. Fan, Y. Chen, W.W. Lin, Y.N. Lu and J.H. Hu, *J. Phys. Chem. C*, 112 (2008) 472.
[49] J. Yang and L. Ngaw, *Top. Catal.*, 55 (2012) 663.
[50] S. Jin, W.Z. Qian, Y. Liu, F. Wei, D.Z. Wang and J.C. Zhang, *Aust. J. Chem.*, 63 (2010) 131.
[51] R.F. Nie, J.H. Wang, L.N. Wang, Y. Qin, P. Chen and Z.Y. Hou, *Carbon*, 50 (2012) 586.
[52] B.Q. Sheng, L. Hu, T.T. Yu, X.Q. Cao and H.W. Gu, *Rsc Adv.*, 2 (2012) 5520.
[53] H.Y. Cheng, X.C. Meng, Q. Wang, J. Ming, Y.C. Yu and F.Y. Zhao, *J. Colloid Interface Sci.*, 377 (2012) 322.

[54] H.Y. Cheng, C.Y. Xi, X.C. Meng, Y.F. Hao, Y.C. Yu and F.Y. Zhao, *J. Colloid Interface Sci.*, 336 (2009) 675.
[55] S. Shironita, T. Takasaki, T. Kamegawa, K. Mori and H. Yamashita, *Catal. Lett.*, 129 (2009) 404.
[56] S. Shironita, T. Takasaki, T. Kamegawa, K. Mori and H. Yamashita, *Top. Catal.*, 53 (2010) 218.
[57] J.X. Pan, J. Liu, S.P. Guo and Z.Y. Yang, *Catal. Lett.*, 131 (2009) 179.
[58] Y. Motoyama, K. Kamo and H. Nagashima, *Org. Lett.*, 11 (2009) 1345.
[59] K.L. Xu, Y. Zhang, X.R. Chen, L. Huang, R. Zhang and J. Huang, *Adv. Synth. Catal.*, 353 (2011) 1260.
[60] K. Parida, G.B.B. Varadwaj, S. Sahu and P.C. Sahoo, *Ind. Eng. Chem. Res.*, 50 (2011) 7849.
[61] A. Indra, P.R. Rajamohanan, C.S. Gopinath, S. Bhaduri and G.K. Lahiri, *Appl. Catal. A: Gen.*, 399 (2011) 117.
[62] S.J. Tauster, *Acc. Chem. Res.*, 20 (1987) 389.
[63] A.M. Stratz, In: Catalysis of Organic Reactions; Kosak J. R. Ed.; Marcel Dekker: New York, (1984) 335.
[64] D.K. Yi, S.S. Lee and J.Y. Ying, *Chem. Mater.*, 18 (2006) 2459.
[65] Y.Q. Lang, Q.Q. Wang, J.M. Xing, B. Zhang and H.Z. Liu, *AlChE J.*, 54 (2008) 2303.
[66] F.W. Zhang, J. Jin, X. Zhong, S.W. Li, J.R. Niu, R. Li and J.T. Ma, *Green Chem.*, 13 (2011) 1238.
[67] H.Q. Liu, M.H. Liang, C. Xiao, N. Zheng, X.H. Feng, Y. Liu, J.L. Xie and Y. Wang, *J. Mol. Catal. A: Chem.*, 308 (2009) 79.
[68] R.Z. Zhang, J.M. Liu, F.W. Li, S.F. Wang, C.G. Xia and W. Sun, *Chin. J. Chem.*, 29 (2011) 525.
[69] Y.F. Liu, A.M. Huang, J. Chen, L. Chen and J.D. Hua, *J. Appl. Polym. Sci.*, 89 (2003) 3661.
[70] V. Kratky, M. Kralik, M. Mecarova, M. Stolcova, L. Zalibera and M. Hronec, *Appl. Catal. A: Gen.*, 235 (2002) 225.
[71] H.B. Ji, Q.P. Long, Y.B. He and X.D. Yao, *Sci. China Chem.*, 53 (2010) 1520.
[72] H.P. Hemantha and V.V. Sureshbabu, *Org. Biomol. Chem.*, 9 (2011) 2597.
[73] H. Wu, L.M. Zhuo, Q. He, X.P. Liao and B. Shi, *Appl. Catal. A: Gen.*, 366 (2009) 44.
[74] A.V. Biradar, A.A. Biradar and T. Asefa, *Langmuir*, 27 (2011) 14408.
[75] V. Udayakumar, S. Alexander, V. Gayathri, Shivakumaraiah and B. Viswanathan, *React. Kinet. Mech. Cat.*, 103 (2011) 341.

[76] S.G. Xu, X.L. Xi, J. Shi and S.K. Cao, *J. Mol. Catal. A: Chem.*, 160 (2000) 287.
[77] X.L. Xi, Y.L. Liu, J. Shi and S.K. Cao, *J. Mol. Catal. A: Chem.*, 192 (2003) 1.
[78] J.L. Pellegatta, C. Blandy, R. Choukroun, C. Lorber, B. Chaudret, P. Lecante and E. Snoeck, *New J. Chem.*, 27 (2003) 1528.
[79] F.A. Harraz, S.E. El-Hout, H.M. Killa and I.A. Ibrahim, *J. Catal.*, 286 (2012) 184.
[80] C.H. Liang, J.G. Han, K.H. Shen, L.G. Wang, D.F. Zhao and H.S. Freeman, *Chem. Eng. J.*, 165 (2010) 709.
[81] P. Sangeetha, P. Seetharamulu, K. Shanthi, S. Narayanan and K.S.R. Rao, *J. Mol. Catal. A: Chem.*, 273 (2007) 244.
[82] P. Sangeetha, K. Shanthi, K.S.R. Rao, B. Viswanathan and P. Selvam, *Appl. Catal. A: Gen.*, 353 (2009) 160.
[83] M.L. Kantam, R. Chakravarti, U. Pal, B. Sreedhar and S. Bhargava, *Adv. Synth. Catal.*, 350 (2008) 822.
[84] L.M. Sikhwivhilu, N.J. Coville, B.M. Pulimaddi, J. Venkatreddy and V. Vishwanathan, *Catal. Commun.*, 8 (2007) 1999.
[85] N. Maity, P.R. Rajamohanan, S. Ganapathy, C.S. Gopinath, S. Bhaduri and G.K. Lahiri, *J. Phys. Chem. C*, 112 (2008) 9428.
[86] F. Chang, H. Kim, B. Lee, S. Park and J. Park, *Tetrahedron Lett.*, 51 (2010) 4250.
[87] V. Pandarus, R. Ciriminna, F. Beland and M. Pagliaro, *Catal. Sci. Technol.*, 1 (2011) 1616.
[88] Y.F. Wang, A.V. Biradar and T. Asefa, *Chemsuschem*, 5 (2012) 132.
[89] Q. Xu, X.M. Liu, J.R. Chen, R.X. Li and X.J. Li, *J. Mol. Catal. A: Chem.*, 260 (2006) 299.
[90] V. Vishwanathan, V. Jayasri, P.M. Basha, N. Mahata, L. Sikhwivhilu and N.J. Coville, *Catal. Commun.*, 9 (2008) 453.
[91] Obraztsova, II, N.K. Eremenko and Y.N. Velyakina, *Kinet. Catal.*, 49 (2008) 401.
[92] C.J. Cheng, X.Y. Wang, L.X. Xing, B. Liu, R. Zhu and Y.F. Hu, *Adv. Synth. Catal.*, 349 (2007) 1775.
[93] A. Mori, T. Mizusaki, M. Kawase, T. Maegawa, Y. Monguchi, S. Takao, Y. Takagi and H. Sajiki, *Adv. Synth. Catal.*, 350 (2008) 406.
[94] S. Rana and K.M. Parida, *Catal. Sci. Technol.*, 2 (2012) 979.
[95] H. Sajiki, K. Hattori and K. Hirota, *J. Org. Chem.*, 63 (1998) 7990.
[96] X.L. Yang, H.F. Liu and H. Zhong, *J. Mol. Catal. A: Chem.*, 147 (1999) 55.

[97] F. Cardenas-Lizana, S. Gomez-Quero, A. Hugon, L. Delannoy, C. Louis and M.A. Keane, *J. Catal.*, 262 (2009) 235.
[98] K. Chary and C. Srikanth, *Catal. Lett.*, 128 (2009) 164.
[99] C. Antonetti, M. Oubenali, A.M.R. Galletti, P. Serp and G. Vannucci, *Appl. Catal. A: Gen.*, 421 (2012) 99.
[100] S.L. Zhao, H.D. Liang and Y.F. Zhou, *Catal. Commun.*, 8 (2007) 1305.
[101] B.J. Zuo, Y. Wang, Q.L. Wang, J.L. Zhang, N.Z. Wu, L.D. Peng, L.L. Gui, X.D. Wang, R.M. Wang and D.P. Yu, *J. Catal.*, 222 (2004) 493.
[102] Y. Wang, J.L. Zhang, X.D. Wang, J.W. Ren, B.J. Zuo and Y.Q. Tang, *Top. Catal.*, 35 (2005) 35.
[103] M. Oubenali, G. Vanucci, B. Machado, M. Kacimi, M. Ziyad, J. Faria, A. Raspolli-Galetti and P. Serp, *Chemsuschem*, 4 (2011) 950.
[104] M. Pietrowski and M. Wojciechowska, *Catal. Today*, 142 (2009) 211.
[105] J. Ning, J. Xu, J. Liu, H. Miao, H. Ma, C. Chen, X. Li, L. Zhou and W. Yu, *Catal. Commun.*, 8 (2007) 1763.
[106] M. Pietrowski, M. Zielinski and M. Wojciechowska, *Catal. Lett.*, 128 (2009) 31.
[107] A. Tijani, B. Coq and F. Figueras, *Appl. Catal.*, 76 (1991) 255.
[108] J. Hou, Y.H. Ma, Y.H. Li, F. Guo and L.H. Lu, *Chem. Lett.*, 37 (2008) 974.
[109] P. Claus, *Appl. Catal. A: Gen.*, 291 (2005) 222.
[110] J.E. Bailie and G.J. Hutchings, *Chem. Commun.*, (1999) 2151.
[111] M. Boronat, P. Concepcion, A. Corma, S. Gonzalez, F. Illas and P. Serna, *J. Am. Chem. Soc.*, 129 (2007) 16230.
[112] F. Cardenas-Lizana, S. Gomez-Quero and M.A. Keane, *Chemsuschem*, 1 (2008) 215.
[113] F. Cardenas-Lizana, S. Gomez-Quero, L. Kiwi-Minsker and M.A. Keane, *Int. J. Nanotechnol.*, 9 (2012) 92.
[114] Y.Y. Chen, J.S. Qiu, X.K. Wang and J.H. Xiu, *J. Catal.*, 242 (2006) 227.
[115] X.Y. Tan, Z.X. Zhang, Z.H. Xiao, Q. Xu, C.H. Liang and X.K. Wang, *Catal. Lett.*, 142 (2012) 788.
[116] K. Shimizu, Y. Miyamoto, T. Kawasaki, T. Tanji, Y. Tai and A. Satsuma, *J. Phys. Chem. C*, 113 (2009) 17803.
[117] F. Cardenas-Lizana, S. Gomez-Quero and M.A. Keane, *Catal. Commun.*, 9 (2008) 475.
[118] D.P. He, H. Shi, Y. Wu and B.Q. Xu, *Green Chem.*, 9 (2007) 849.
[119] L.Q. Liu, B.T. Qiao, Y.B. Ma, J. Zhang and Y.Q. Deng, *Dalton Trans.*, (2008) 2542.

[120] F. Cardenas-Lizana, D. Lamey, N. Perret, S. Gomez-Quero, L. Kiwi-Minsker and M.A. Keane, *Catal. Commun.*, 21 (2012) 46.
[121] N. Perret, X.D. Wang, L. Delannoy, C. Potvin, C. Louis and M.A. Keane, *J. Catal.*, 286 (2012) 172.
[122] A. Corma, C. Gonzalez-Arellano, M. Iglesias and F. Sanchez, *Appl. Catal. A: Gen.*, 356 (2009) 99.
[123] A. Corma and P. Serna, *Nat. Protoc.*, 1 (2006) 2590.
[124] P. Serna, M. Boronat and A. Corma, *Top. Catal.*, 54 (2011) 439.
[125] E.A. Gelder, S.D. Jackson and C.M. Lok, Competitive hydrogenation of nitrobenzene, nitrosobenzene and azobenzene, In: Catalysis of Organic Reactions; S.R. Schmidt Ed.; 115 (2007) 167.
[126] G. Richner, J.A. van Bokhoven, Y.M. Neuhold, M. Makosch and K. Hungerbuhler, *Phys. Chem. Chem. Phys.*, 13 (2011) 12463.
[127] M. Makosch, J. Sa, C. Kartusch, G. Richner, J.A. van Bokhoven and K. Hungerbuhler, *Chemcatchem*, 4 (2012) 59.
[128] F. Cardenas-Lizana, S. Gomez-Quero, N. Perret and M.A. Keane, *Catal. Sci. Technol.*, 1 (2011) 652.
[129] F. Cardenas-Lizana, S. Gomez-Quero, N. Perret and M.A. Keane, *Gold Bull.*, 42 (2009) 124.
[130] F. Cardenas-Lizana, S. Gomez-Quero, H. Idriss and M.A. Keane, *J. Catal.*, 268 (2009) 223.
[131] J.Q. Lu, J.J. Bravo-Suarez, A. Takahashi, M. Haruta and S.T. Oyama, *J. Catal.*, 232 (2005) 85.
[132] D.J. Guo and H.L. Li, *Carbon*, 43 (2005) 1259.
[133] Y.Y. Chen, C.A. Wang, H.Y. Liu, J.S. Qiu and X.H. Bao, *Chem. Commun.*, (2005) 5298.
[134] B.J. Li, H.B. Li and Z. Xu, *J. Phys. Chem. C*, 113 (2009) 21526.
[135] K. Shimizu, Y. Miyamoto and A. Satsuma, *J. Catal.*, 270 (2010) 86.
[136] Y. Mikami, A. Noujima, T. Mitsudome, T. Mizugaki, K. Jitsukawa and K. Kaneda, *Chem. Lett.*, 39 (2010) 223.
[137] E. Auer, A. Freund, M. Gross, R. Hartung and P. Panster, New catalytic systems for the selective hydrogenation of halogenated aromatic nitro compounds, In Catalysis of Organic Reactions; F.E. Herkes Ed.; 75, 1998, 551.
[138] G.Y. Fan, L. Zhang, H.Y. Fu, M.L. Yuan, R.X. Li, H. Chen and X.J. Li, *Catal. Commun.*, 11 (2010) 451.
[139] W. Pascoe, In: Catalysis of Organic Reactions; Rylander P.N.; Greenfield H.; Augustine R.L. Ed.; Marcel Dekker: New York, (1988) 121.

[140] W.P. Dunworth and F.F. Nord, *J. Am. Chem. Soc.*, 74 (1952) 1459.
[141] H.B. Lu, H.B. Yin, Y.M. Liu, T.S. Jiang and L.B. Yu, *Catal. Commun.*, 10 (2008) 313.
[142] J.F. Ding, L.G. Chen, R. Shao, J. Wu and W.T. Dong, *React. Kinet. Mech. Cat.*, 106 (2012) 225.
[143] A. Nieto-Marquez, S. Gil, A. Romero, J.L. Valverde, S. Gomez-Quero and M.A. Keane, *Appl. Catal. A: Gen.*, 363 (2009) 188.
[144] J. Xiong, J.X. Chen and J.Y. Zhang, *Catal. Commun.*, 8 (2007) 345.
[145] N. Yao, J.X. Chen, J.X. Zhang and J.Y. Zhang, *Catal. Commun.*, 9 (2008) 1510.
[146] J.X. Chen, N. Yao, R.J. Wang and J.Y. Zhang, *Chem. Eng. J.*, 148 (2009) 164.
[147] N. Mahata, A.F. Cunha, J.J.M. Orfao and J.L. Figueiredo, *Appl. Catal. A: Gen.*, 351 (2008) 204.
[148] N. Mahata, A.F. Cunha, J.J.M. Orfao and J.L. Figueiredo, *Catal. Commun.*, 10 (2009) 1203.
[149] N. Mahata, A.F. Cunha, J.J.M. Orfao and J.L. Figueiredo, *Chemcatchem*, 2 (2010) 330.
[150] Z.J. Wu, J. Chen, Q. Di and M.H. Zhang, *Catal. Commun.*, 18 (2012) 55.
[151] J.H. Wang, Z.L. Yuan, R.F. Nie, Z.Y. Hou and X.M. Zheng, *Ind. Eng. Chem. Res.*, 49 (2010) 4664.
[152] C. Wang, J.S. Qiu, C.H. Liang, L. Xing and X.M. Yang, *Catal. Commun.*, 9 (2008) 1749.
[153] J. Wang, G.L. Fan, H. Wang and F. Li, *Ind. Eng. Chem. Res.*, 50 (2011) 13717.
[154] Y. Du, H.L. Chen, R.Z. Chen and N.P. Xu, *Appl. Catal. A: Gen.*, 277 (2004) 259.
[155] A.L. Wang, H.B. Yin, M. Ren, H.H. Lu, J.J. Xue and T.S. Jiang, *New J. Chem.*, 34 (2010) 708.
[156] J. Relvas, R. Andrade, F.G. Freire, F. Lemos, P. Araujo, M.J. Pinho, C.P. Nunes and F.R. Ribeiro, *Catal. Today*, 133 (2008) 828.
[157] Y.F. Zheng, K. Ma, H.L. Wang, X. Sun, J. Jiang, C.F. Wang, R. Li and J.T. Ma, *Catal. Lett.*, 124 (2008) 268.
[158] R.Z. Chen, Q.Q. Wang, Y. Du, W.H. Xing and N.P. Xu, *Chem. Eng. J.*, 145 (2009) 371.
[159] W. Lin, H. Cheng, J. Ming, Y. Yu and F. Zhao, *J. Catal.*, 291 (2012) 149.
[160] V. Mohan, C.V. Pramod, M. Suresh, K.H.P. Reddy, B.D. Raju and K.S.R. Rao, *Catal. Commun.*, 18 (2012) 89.

[161] L. Zhao, J.X. Chen and J.Y. Zhang, *J. Mol. Catal. A: Chem.*, 246 (2006) 140.
[162] H. Li, H. Lin, S.H. Xie, W.L. Dai, M.H. Qiao, Y.F. Lu and H.X. Li, *Chem. Mater.*, 20 (2008) 3936.
[163] R. Raja, V.B. Golovko, J.M. Thomas, A. Berenguer-Murcia, W.Z. Zhou, S.H. Xie and B.F.G. Johnson, *Chem. Commun.*, (2005) 2026.
[164] L. Xing, J.S. Qiu, C.H. Liang, C. Wang and L. Mao, *J. Catal.*, 250 (2007) 369.
[165] F. Cardenas-Lizana, S. Gomez-Quero and M. Keane, *Catal. Lett.*, 127 (2009) 25.
[166] F. Cardenas-Lizana, S. Gomez-Quero, C.J. Baddeley and M.A. Keane, *Appl. Catal. A: Gen.*, 387 (2010) 155.
[167] F. Cardenas-Lizana, S. Gomez-Quero, G. Jacobs, Y.Y. Ji, B.H. Davis, L. Kiwi-Minsker and M.A. Keane, *J. Phys. Chem. C*, 116 (2012) 11166.
[168] T. Joseph, K.V. Kumar, A.V. Ramaswamy and S.B. Halligudi, *Catal. Commun.*, 8 (2007) 629.
[169] P. Lu, T. Teranishi, K. Asakura, M. Miyake and N. Toshima, *J. Phys. Chem. B*, 103 (1999) 9673.
[170] P. Lu and N. Toshima, *Bull. Chem. Soc. Jpn.*, 73 (2000) 751.
[171] Y. Zhang, F. Hou and Y.W. Tan, *Chem. Commun.*, 48 (2012) 2391.
[172] X.X. Han, R.X. Zhou, G.H. Lai, B.H. Yue and X.M. Zheng, *J. Mol. Catal. A: Chem.*, 209 (2004) 83.
[173] X.X. Han, R.X. Zhou, G.H. Lai and X.M. Zheng, *React. Kinet. Catal. Lett.*, 83 (2004) 55.
[174] M.H. Liu, J. Zhang, J.Q. Liu and W.W. Yu, *J. Catal.*, 278 (2011) 1.
[175] K.N. Rao, B.M. Reddy and S.E. Park, *Catal. Commun.*, 11 (2009) 142.
[176] F. Cardenas-Lizana, B. Bridier, C.C.K. Shin, J. Perez-Ramirez and L. Kiwi-Minsker, *Chemcatchem*, 4 (2012) 668.
[177] M. Boronat and A. Corma, *Langmuir*, 26 (2010) 16607.
[178] R.X. Liu, Y.C. Yu, K. Yoshida, G.M. Li, H.X. Jiang, M.H. Zhang, F.Y. Zhao, S. Fujita and M. Arai, *J. Catal.*, 269 (2010) 191.
[179] W. Leitner, *Green Chem.*, 9 (2007) 923.
[180] R.A. Sheldon, *Green Chem.*, 10 (2008) 359.
[181] C.J. Li, *Green Chem.*, 10 (2008) 151.
[182] J. Li, X.Y. Shi, Y.Y. Bi, J.F. Wei and Z.G. Chen, *Acs Catal.*, 1 (2011) 657.
[183] Z.Y. Sun, H.Y. Zhang, G.M. An, G.Y. Yang and Z.M. Liu, *J. Mater. Chem.*, 20 (2010) 1947.

[184] S. Nishimura, Handbook of Heterogeneous Catalytic Hydrogenation for Organic Synthesis, John Wiley and Sons, New York, 2001.
[185] C. Reichardt, *Org. Process Res. Dev.*, 11 (2007) 105.
[186] C.X. Xiao, H.Z. Wang, X.D. Mu and Y. Kou, *J. Catal.*, 250 (2007) 25.
[187] D.Q. Xu, Z.Y. Hu, W.W. Li, S.P. Luo and Z.Y. Xu, *J. Mol. Catal. A: Chem.*, 235 (2005) 137.
[188] S. Saleh, E. Fayad, M. Azouri, J.C. Hierso, J. Andrieu and M. Picquet, *Adv. Synth. Catal.*, 351 (2009) 1621.
[189] P.G. Jessop, *Green Chem.*, 13 (2011) 1391.
[190] C.A. Eckert, B.L. Knutson and P.G. Debenedetti, *Nature*, 383 (1996) 313.
[191] A. Baiker, *Chem. Rev.*, 99 (1999) 453.
[192] B.M. Bhanage and M. Arai, *Catal. Rev. Sci. Eng.*, 43 (2001) 315.
[193] J.R. Hyde, P. Licence, D. Carter and M. Poliakoff, *Appl. Catal. A: Gen.*, 222 (2001) 119.
[194] W. Leitner, *Acc. Chem. Res.*, 35 (2002) 746.
[195] E.J. Beckman, *J. Supercrit. Fluids*, 28 (2004) 121.
[196] Y.F. Hao, R.X. Liu, X.C. Meng, H.Y. Cheng and F.Y. Zhao, *J. Mol. Catal. A: Chem.*, 335 (2011) 183.
[197] D. Ferri, T. Burgi and A. Baiker, *Phys. Chem. Chem. Phys.*, 4 (2002) 2667.
[198] M. Burgener, D. Ferri, J.D. Grunwaldt, T. Mallat and A. Baiker, *J. Phys. Chem. B*, 109 (2005) 16794.
[199] I. Fujita, T. Yamada, Y. Akiyama, H.Y. Cheng, F.Y. Zhao and M. Arai, *J. Supercrit. Fluids*, 54 (2010) 190.
[200] P.G. Jessop and B. Subramaniam, *Chem. Rev.*, 107 (2007) 2666.
[201] G.R. Akien and M. Poliakoff, *Green Chem.*, 11 (2009) 1083.
[202] S. Narayan, J. Muldoon, M.G. Finn, V.V. Fokin, H.C. Kolb and K.B. Sharpless, *Angew. Chem. Int. Ed.*, 44 (2005) 3275.
[203] H.C. Hailes, *Org. Process Res. Dev.*, 11 (2007) 114.
[204] A.A. Deshmukh, A.K. Prashar, A.K. Kinage, R. Kumar and R. Meijboom, *Ind. Eng. Chem. Res.*, 49 (2010) 12180.
[205] Y.Z. Hao, Z.X. Li and J.L. Tian, *J. Mol. Catal. A: Chem.*, 265 (2007) 258.
[206] X. Huang, Y.P. Wang, X.P. Liao and B. Shi, *Chem. Commun.*, (2009) 4687.
[207] P. Maity, S. Basu, S. Bhaduri and G.K. Lahiri, *Adv. Synth. Catal.*, 349 (2007) 1955.
[208] B. Sreedhar, D.K. Devi and D. Yada, *Catal. Commun.*, 12 (2011) 1009.

[209] S. Samanta and A.K. Nandi, *J. Phys. Chem. C*, 113 (2009) 4721.
[210] Y. Hu, Y.Y. Yu, X.G. Zhao, H.M. Yang, B. Feng, H.A. Li, Y.X. Qiao, L. Hua, Z.Y. Pan and Z.S. Hou, *Sci. China Chem.*, 53 (2010) 1541.
[211] M. Jorge and M. Cordeiro, *J. Phys. Chem. C*, 111 (2007) 17612.
[212] X.B. Lou, L. He, Y. Qian, Y.M. Liu, Y. Cao and K.N. Fan, *Adv. Synth. Catal.*, 353 (2011) 281.
[213] Y. Jang, S. Kim, S.W. Jun, B.H. Kim, S. Hwang, I.K. Song, B.M. Kim and T. Hyeon, *Chem. Commun.*, 47 (2011) 3601.
[214] Y.J. Gao, D. Ma, C.L. Wang, J. Guan and X.H. Bao, *Chem. Commun.*, 47 (2011) 2432.
[215] Z.Y. Ji, X.P. Shen, G.X. Zhu, H. Zhou and A.H. Yuan, *J. Mater. Chem.*, 22 (2012) 3471.
[216] H.K. Kadam and S.G. Tilve, *Rsc Adv.*, 2 (2012) 6057.
[217] X.N. Li, J.H. Zhang, Y.Z. Xiang, L. Ma, Q.F. Zhang, C.S. Lu, H. Wang and Y. Bai, *Sci. China Ser. B*, 51 (2008) 248.
[218] L. Xu, X.N. Li, Y.F. Zhu and Y.Z. Xiang, *New J. Chem.*, 33 (2009) 2051.
[219] Y.Z. Xiang, X.N. Li, C.S. Lu, L. Ma and Q.F. Zhang, *Appl. Catal. A: Gen.*, 375 (2010) 289.
[220] L. Zhou, H.Z. Gu and X.H. Yan, *Catal. Lett.*, 132 (2009) 16.
[221] M. Pietrowski, *Green Chem.*, 13 (2011) 1633.
[222] N. Garcia, P. Garcia-Garcia, M.A. Fernandez-Rodriguez, R. Rubio, M.R. Pedrosa, F.J. Arnaiz and R. Sanz, *Adv. Synth. Catal.*, 354 (2012) 321.
[223] K.H.P. Reddy, R. Rahul, S.S.V. Reddy, B.D. Raju and K.S.R. Rao, *Catal. Commun.*, 10 (2009) 879.
[224] A.L. Sun, Z.F. Qin and J.G. Wang, *Catal. Lett.*, 79 (2002) 33.
[225] K. Nomura, M. Ishino and M. Hazama, *J. Mol. Catal.*, 66 (1991) L1.
[226] C. Longo, J. Alvarez, M. Fernandez, A.J. Pardey, S.A. Moya, P. Baricelli and M.M. Mdleleni, *Polyhedron*, 19 (2000) 487.
[227] A.M. Tafesh and M. Beller, *Tetrahedron Lett.*, 36 (1995) 9305.
[228] D. Andreeva, V. Idakiev, T. Tabakova and A. Andreev, *J. Catal.*, 158 (1996) 354.
[229] O. Thinon, F. Diehl, P. Avenier and Y. Schuurman, *Catal. Today*, 137 (2008) 29.
[230] L.Q. Liu, B.T. Qiao, Z.J. Chen, J. Zhang and Y.Q. Deng, *Chem. Commun.*, (2009) 653.
[231] L. He, L.C. Wang, H. Sun, J. Ni, Y. Cao, H.Y. He and K.N. Ean, *Angew. Chem. Int. Ed.*, 48 (2009) 9538.

[232] X.Z. Liu and S.W. Lu, *J. Mol. Catal. A: Chem.*, 300 (2009) 36.
[233] M. Vigano, F. Ragaini, M.G. Buonomenna, R. Lariccia, A. Caselli, E. Gallo, S. Cenini, J.C. Jansen and E. Drioli, *Chemcatchem*, 2 (2010) 1150.
[234] O. Rios-Berny, S.O. Flores, I. Cordova and M.A. Valenzuela, *Tetrahedron Lett.*, 51 (2010) 2730.
[235] T.Y. Zhang, L.Y. You and Y.L. Zhang, *Dyes Pigments*, 68 (2006) 95.
[236] F. Mahdavi, T.C. Bruton and Y.Z. Li, *J. Org. Chem.*, 58 (1993) 744.
[237] V. Brezova, A. Blazkova, I. Surina and B. Havlinova, J. Photochem. Photobiol. A, 107 (1997) 233.
[238] A. Maldotti, L. Andreotti, A. Molinari, S. Tollari, A. Penoni and S. Cenini, *J. Photochem. Photobiol. A*, 133 (2000) 129.
[239] J.L. Ferry and W.H. Glaze, *Langmuir*, 14 (1998) 3551.
[240] S.O. Flores, O. Rios-Bernij, M.A. Valenzuela, I. Cordova, R. Gomez and R. Gutierrez, *Top. Catal.*, 44 (2007) 507.
[241] H. Tada, T. Ishida, A. Takao and S. Ito, *Langmuir*, 20 (2004) 7898.
[242] H.Q. Wang, J.P. Yan, Z.M. Zhang and W.F. Chang, *React. Kinet. Catal. Lett.*, 97 (2009) 91.
[243] H.Q. Wang, J.P. Yan, W.F. Chang and Z.M. Zhang, *Catal. Commun.*, 10 (2009) 989.
[244] W.Y. Ahn, S.A. Sheeley, T. Rajh and D.M. Cropek, *Appl. Catal. B: Environ.*, 74 (2007) 103.
[245] H.Y. Huang, J.H. Zhou, H.L. Liu, Y.H. Zhou and Y.Y. Feng, *J. Hazard. Mater.*, 178 (2010) 994.
[246] W.M. Wu, G.D. Liu, S.J. Liang, Y. Chen, L.J. Shen, H.R. Zheng, R.S. Yuan, Y.D. Hou and L. Wu, *J. Catal.*, 290 (2012) 13.
[247] H. Kominami, S.I. Iwasaki, T. Maeda, K. Imamura, K. Hashimoto, Y. Kera and B. Ohtani, *Chem. Lett.*, 38 (2009) 410.
[248] K. Imamura, S. Iwasaki, T. Maeda, K. Hashimoto, B. Ohtani and H. Kominami, *Phys. Chem. Chem. Phys.*, 13 (2011) 5114.
[249] K. Imamura, K. Hashimoto and H. Kominami, *Chem. Commun.*, 48 (2012) 4356.
[250] M. Arai, A. Obata and Y. Nishiyama, *React. Kinet. Catal. Lett.*, 61 (1997) 275.
[251] L.L. Santos, P. Serna and A. Corma, *Chem. Eur. J.*, 15 (2009) 8196.
[252] D.B. Bagal, R.A. Watile, M.V. Khedkar, K.P. Dhake and B.M. Bhanage, *Catal. Sci. Technol.*, 2 (2012) 354.
[253] Y.F. Zheng, K. Ma, H. Li, J. Li, J.G. He, X. Sun, R. Li and J.M. Ma, *Catal. Lett.*, 128 (2009) 465.

[254] V. Macho, M. Kralik, J. Hudec and J. Cingelova, *J. Mol. Catal. A: Chem.*, 209 (2004) 69.
[255] J. Huang, L. Yu, L. He, Y.M. Liu, Y. Cao and K.N. Fan, *Green Chem.*, 13 (2011) 2672.
[256] Y.Z. Xiang, Q.Q. Meng, X.N. Li and J.G. Wang, *Chem. Commun.*, 46 (2010) 5918.
[257] L. Hu, X.Q. Cao, D.H. Ge, H.Y. Hong, Z.Q. Guo, L. Chen, X.H. Sun, J.X. Tang, J.W. Zheng, J.M. Lu and H.W. Gu, *Chem. Eur. J.*, 17 (2011) 14283.
[258] C. del Pozo, A. Corma, M. Iglesias and F. Sánchez, *J. Catal.*, 291 (2012) 110.
[259] M.M. Dell'Anna, P. Mastrorilli, A. Rizzuti and C. Leonelli, *Appl. Catal. A: Gen.*, 401 (2011) 134.
[260] Y. Yamane, X.H. Liu, A. Hamasaki, T. Ishida, M. Haruta, T. Yokoyama and M. Tokunaga, *Org. Lett.*, 11 (2009) 5162.
[261] A. Zanardi, J.A. Mata and E. Peris, *Chem. Eur. J.*, 16 (2010) 10502.
[262] Y. Liu, W. Chen, C. Feng and G.J. Deng, *Chem. Asian J.*, 6 (2011) 1142.
[263] X.J. Cui, Y. Zhang, F. Shi and Y.Q. Deng, *Chem. Eur. J.*, 17 (2011) 2587.
[264] C.C. Lee and S.T. Liu, *Chem. Commun.*, 47 (2011) 6981.
[265] C.H. Tang, L. He, Y.M. Liu, Y. Cao, H.Y. He and K.N. Fan, *Chem. Eur. J.*, 17 (2011) 7172.
[266] T. Ikawa, Y. Fujita, T. Mizusaki, S. Betsuin, H. Takamatsu, T. Maegawa, Y. Monguchi and H. Sajiki, *Org. Biomol. Chem.*, 10 (2012) 293.

In: Aniline
Editors: K. Hernandez, M. Holloway
ISBN: 978-1-62417-417-9

Chapter 2

ANILINE-BASED POLYBENZOXAZINE AND THEIR COPOLYMERS OR COMPOSITES: MOLECULAR DESIGN AND POTENTIAL APPLICATIONS

***Chanchira Jubsilp*[1,*], *Sarawut Rimdusit*[2] *and Tsutomu Takeichi*[3]**

[1] Department of Chemical Engineering, Srinakharinwirot University, Ongkharak, Nakhonnayok, Thailand
[2] Polymer Engineering Laboratory, Department of Chemical Engineering, Chulalongkorn University, *Patumwan*, *Bangkok*, Thailand
[3] Department of Environmental and Life Sciences, Toyohashi University of Technology, Tempaku-cho, Toyohashi, Japan

ABSTRACT

Polybenzoxazines as a novel phenolic type thermosetting polymer have been developed to overcome the shortcomings associated with the use of the traditional novolac and resol-type phenolic resins. They can be prepared from cyclic benzoxazines via thermally induced ring-opening polymerization. Various benzoxazine monomers can be synthesized from mono- or biphenols, paraformaldehyde and aniline. Then, the different

[*] Corresponding author Chanchira Jubsilp: e-mail: Chanchira@swu.ac.th.

approaches for performance improvement of polybenzoxazine have been investigated.

These approaches include hybrid with other high performance polymers and reinforcing with natural fillers, long fibers, and inorganic fillers. In these approaches, lowering of the cure temperature, enhancement of toughness, and improvement of thermal and mechanical properties were achieved. In addition, some application potentials of polybenzoxazine-based materials are also considered.

INTRODUCTION

Organic compounds have qualities that make them necessary to sustaining life on earth and include complex molecules such as genetic information bearing deoxyribonucleic acid (DNA) and proteins that constitute essential compounds of our blood, muscles and skin.

Organic chemicals identified by functional groups appear in the important areas of applications like clothing, fuels, dyes, medicines, agro-chemicals, polymers etc. The functional groups are defined as an atom or group of atoms joined in a specific manner which is responsible for the characteristic chemical properties of the organic compounds such as hydroxyl group (−OH), carboxylic acid group (−COOH), aldehyde group (−CHO), and amine group ($-NH_2$) etc. Nitrogen atoms-based functional groups impart physicochemical characteristics to these molecules and respond for their unique chemical reactivity patterns. There are many functional groups, which contain one or more nitrogen atoms.

Some categories of compounds based on these functional groups include nitrocompounds, amines, cyanides, isocyanides and diazo compounds. In case of amine compounds, the main commercial uses are found as intermediates in the synthesis of dyes, synthetic fibers, polymer, and pharmaceuticals. Commercially amines are prepared from a variety of compounds; generally, they are synthesized by reduction of nitrocompounds by catalytic hydrogenation or by using a metal and acid. Aniline manufactured by the reduction of nitrobenzene is one of the most important of all amines.

Aniline, phenylamine or aminobenzene, the prototypical aromatic amine, is an organic compound with the formula $C_6H_5NH_2$ consisting of an amine nitrogen bonded to a benzene ring. In recent years, anilines are still concentrated to largely utilize for many industrial chemicals as diphenylmethane diisocyanate (MDI) which is a precursor to polyurethane, rubber ingredients, organic and natural pigments, cyclohexylamines, and pesticides.

In the MDI process, aniline is condensed with formaldehyde to produce methylene dianiline (MDA), which in turn reacted with phosgene to form MDI (BASF Corporation, 2007). Recently, aniline is used as a precursor of benzoxazine resins which are polymerized by thermal ring-opening of cyclic monomer to polybenzoxazine, a novel type of high performance thermosetting phenolic resin (Ishida and Agag, 2011).

Benzoxazine resins can be readily prepared by a combination of a phenolic derivative, paraformaldehyde, and a primary amine. Small molecular weight benzoxazine resins were firstly prepared and reported by Holly and Cope (1944).

The mono-functional benzoxazine resins were synthesized from bis-(3,4-dihydro-2H-1,3-benzoxazine-3-yl)-methylene in solvent by the reaction of 2-hydroxybenzylamine (1 mole) with formaldehyde (2 mol). This benzoxazine resin can be further reacted with 1 mol of phenol to form 3,4-dihydro-3-(2-hydroxy) benzyl-2H-1,3-benzoxazine. However, the polymerization of phenol-based monofunctional benzoxazine resins leads to only the oligomer formation.

Therefore, this approach could not make any polymeric materials since the thermal dissociation of the resin competed with chain propagation reaction so that high molecular weight structures could not be obtained. However, they might be useful as reactive diluent to facilitate processing. Recently, a novel type of bi-functional or multi-functional benzoxazine resins has been developed by Ishida (1996).

The structure of a typical bi-functional polybenzoxazine was synthesized by crosslinking bi-functional benzoxazine resin (BA-a) prepared from bisphenol-A, aniline and paraformaldehyde through a ring-opening polymerization. The resulting polybenzoxazine possesses various outstanding properties such as high processability due to very low melt viscosity before polymerization, near-zero volumetric shrinkage or expansion upon polymerization, high glass transition temperature (T_g), high thermal stability and excellent mechanical properties, low water uptake, and low coefficient of thermal expansion (CTE).

Furthermore, the polybenzoxazines render low dielectric constant and dissipation loss with great molecular design flexibility, thus showing high potential in many applications. Therefore, it is soon realized that the polybenzoxazines can offer various superior properties compared to the conventional phenolic novolac or resole or epoxy resins.

ANILINE-BASED MONO- OR BI-FUNCTIONAL POLYBENZOXAZINES

Polymerization processes of benzoxazine resins prepared from phenols, paraformaldehyde and primary amines into polybenzoxazine differ from the traditional phenolic resins in that the phenolic moieties are connected by a Mannich base bridge (–CH_2–N(R)–CH_2–) instead of a methylene (–CH_2–) bridge associated with the traditional phenolics. The benzoxazine resins, a molecule where an oxazine ring, a heterocyclic six-membered ring with oxygen and nitrogen atom attached to a benzene ring, are capable of undergoing ring-opening polymerization upon heating without the aid of a curing agent (strong acid and alkaline); therefore, no condensation by-products are released during a fabrication process, thus offering an excellent dimensional stability for the final product. In addition, their polymers have been attracted great attention as high performance materials for engineering and structural applications since they possess various interesting properties that combine outstanding high heat resistance and flame retardance of phenolic resins with excellent mechanical performance and molecular design flexibility of advanced epoxy resins (Ishida, 1998). In the past recent years, the properties of the polybenzoxazines have been extensively studied to provide a good molecular understanding of the polymer, as well as to meet the industrial interests. For exemplification, the chemical structures of benzoxazine resins, i.e., mono-functional (phenol and aniline-based benzoxazine resin, Ph-a) and bi-functional (bisphenol-A and aniline-based benzoxazine resin, BA-a) structures are shown in Figure 1.

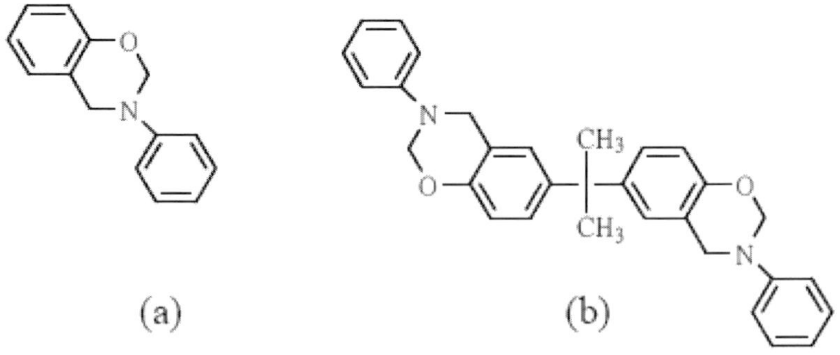

Figure 1. (a) Ph-a benzoxazine resin structure and (b) BA-a benzoxazine resin structure.

Mono-Functional Polybenzoxazine Based on Phenol and Aniline

As above mentioned, polybenzoxazine has many attractive characteristics especially for various highly-filled applications such as electronic packagings, matrix resin for fiber-reinforced polymers, and adhesives. For these applications, the use of a low viscosity polymer is helpful to balance the steep increase of the viscosity as the filler loading is increased. Therefore, the development of mono-functional benzoxazine resins based on unsubstituted phenol, paraformaldehyde and primary aromatic amines, which are liquid at room temperature, has been reported. The polymerization behavior of these resins is shown in Figure 2 and the properties of the resulting thermosetting polymers were also subsequently examined by Wang and Ishida (2002). In a series of arylamine-based mono-fucntional benzoxazine resins, phenol and aniline-based mono-functional benzoxazine resin (Ph-a) showed superior processability as well as thermal stability compared to phenol and toluidine-based mono-functional benzoxazine resin (Ph-mt) or phenol and xylidine-based mono-functional benzoxazine resin (Ph-35x). In addition, degradation temperature at 5% weight loss (T_{d5}) and residual weight at 800°C of the Ph-a polymer were also reported to be 380°C and 38%, respectively. Its $T_{g,DSC}$ of 142°C is the highest among the three mono-functional resins tested ($T_{g,DSC}$ of Ph-mt = 122°C and $T_{g,DSC}$ of Ph-35x = 130°C).

Mono-Functional Polybenzoxazine Based on Cresol and Aniline

Another mono-functional polybenzoxazine was studied by Takeichi et al. (2004). They have involved the preparation of similar mono-functional monomers from the previous study, but replacing phenol by cresol as presented in Figure 3.

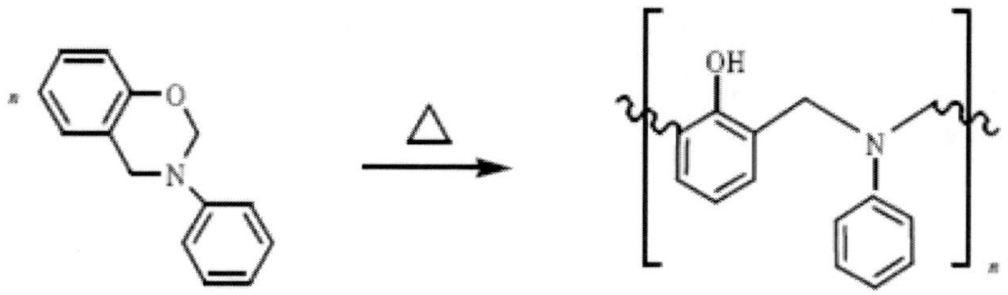

Figure 2. Polymerization of Ph-a benzoxazine resin.

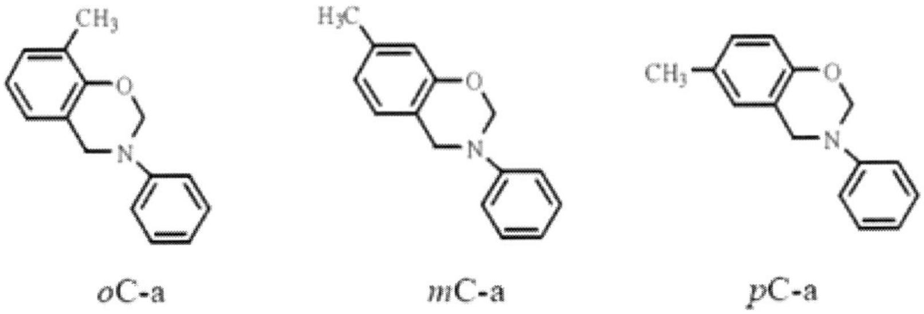

Figure 3. Cresol and aniline-based benzoxazine resins.

Synthesis of a series of cresol-based benzoxazine monomers and the properties of the polymers was prepared from aniline, paraformaldehyde and various cresols i.e. *o*-cresol, *m*-cresol, and *p*-cresol. The softening temperature test was performed to estimate the glass transition temperatures (T_gs) and thermogravimetric analysis (TGA) was carried out to estimate thermal stability of the polybenzoxazines and the results are summarized in Table 1. From the table, the increment in thermal properties, T_g and T_d, of the aniline-based polybenzoxazines from various cresols was in the following order: P*m*C-a > P*p*C-a > P*o*C-a. We can see that P*m*C-a shows the highest T_g value due to both *ortho* and *para*-positions as a reactive site for polymerization *via* electrophilic substitution reaction, resulting in a high degree of polymerization and, hence, high thermal properties while polymerization of the P*p*C-a and the P*o*C-a is possible at the less reactive *para*-position and only at the *ortho*-position, respectively. Therefore, it was confirmed that the ring-opening polymerization of the benzoxazine resins occurred preferentially at the *ortho*-position. However, the ring-opening reaction was found to be possible at the *para*-position as well.

In addition, the thermal curing reaction of the benzoxazine resin based on *p*-cresol, paraformaldehyde and aniline was studied by Huang et al. (2006).

Table 1. Thermal properties of cresol and aniline-based polybenzoxazines

Polybenzoxazine	Softening temperature (°C)	T_{d5} (°C)	T_{d10} (°C)	Residual weight (%)
Poly(*o*C-a)	130	275	310	42
Poly(*m*C-a)	169	280	322	37
Poly(*p*C-a)	164	282	314	37

The model compound 6-methyl-3-phenyl-3,4-dihydro-2H-benzoxazine (PMB or P*p*C-a) was synthesized in their investigation. Autocatalytic model, Kissinger and Ozawa method based on non-isothermal DSC data, have been applied to calculate the kinetic parameters for cure reaction of the PMB. From the results, the average activation energy corresponding to polymerization of the PMB (124.1 kJ/mol) was higher than that of the model compound 3-(3-ethynyl-phenyl)-6-methyl-3,4-dihydro-2H benzoxazine, EPMB, (103.8 kJ/mol). This phenomenon suggested that the polymerization of the PMB probably had lower reactivity than EPMB thus required higher polymerization temperature.

Mono-Functional Polybenzoxazine Based on Cardanol and Aniline

Benzoxazine resins that are liquid at room temperature and can potentially be developed into high performance polybenzoxazines are of interest for various applications, such as fiber-reinforced composites, where higher filler loading (highly-filled systems) and faster polymerization can be achieved [8]. Mono-functional liquid benzoxazine resin from cardanol and aniline (CBO) for bio-composite applications has been investigated by Rao and Palanisamy (2011). A mono-functional liquid benzoxazine resin was prepared by condensation reaction of aniline and formaldehyde with cardanol, a by product of cashew nut shell liquid (CNSL). Synthesis of cardanol-aniline based benzoxazine monomer and its conversion to polybenzoxazine is displayed in Figure 4.

For the CBO resin, differential scanning calorimetry (DSC) thermogram indicated an onset of thermal polymerization of the resin at 230°C and a final cure temperature of 291°C with a peak maximum at 272°C. The heat of polymerization based on the area of an exothermic peak was found to be 78 J/g. Thermogravimetric analysis (TGA) showed a char yield of 24% at 800°C with a maximum decomposition temperature of 432°C. Thermomechanical properties from dynamic mechanical analysis (DMA), i.e. storage modulus at 38°C and glass transition temperature of the cured CBO were also reported to be 2.71×10^{8} Pa and 104.22°C, respectively. Furthermore, the authors had suggested that this liquid cardanol-aniline based benzoxazine resin derived from renewable resource, could be a promising candidate for making bio-based composites as it can be used alone or as a diluent with other benzoxazine resins.

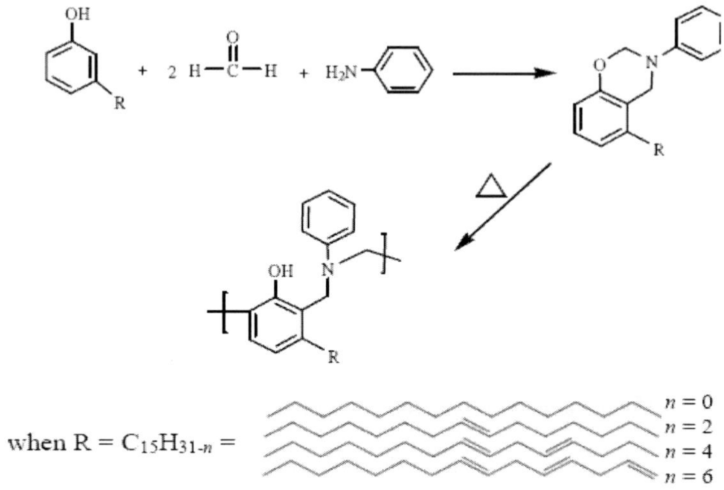

Figure 4. Polymerization reaction of cardanol and aniline-based polybenzoxazine.

Bi-Functional Polybenzoxazines Based on Biphenol and Aniline

According to previous publications, it is difficult to obtain high molecular weight linear polybenzoxazine from mono-functional benzoxazine resin by thermal cure, instead polybenzoxazine derived from bi-functional resin is typically used. Most bi-functional benzoxazine resins are prepared from aromatic biphenols, paraformaldehyde, and mono-amines, especially aniline. The great varieties of an aromatic biphenols and amine allow considerable molecular design flexibility for the benzoxazine resins. Some special functional group can be introduced via biphenols or mono-amines to provide some desired properties such as bi-functional benzoxazine monomer based on bisphenol-A and aniline, BA-a, and polybenzoxazines therefrom are depicted in Figure 5.

Ning and Ishida (1994a, 1994b) reported a superb thermosetting property with simple preparation steps of bi-functional benzoxazine monomers based on bisphenol-A, paraformaldehyde and aniline. The resin is polymerized to cross-linked polybenzoxazine and is expressed as poly(BA-a). The bi-functional benzoxazine monomers were reported to exhibit excellent mechanical and thermal properties with good handling capability for resin processing and composite manufacturing, e.g. the glass transition temperature of 190°C, the degradation temperature ca. 310°C, tensile modulus of 3.2 GPa, and tensile strength of 58 MPa.

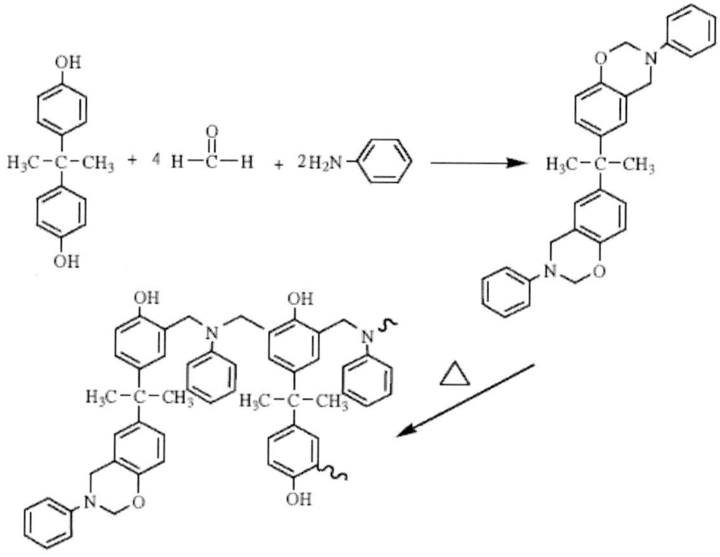

Figure 5. Typical bi-functional BA-a resin and polybenzoxazine therefrom.

In addition, they offered greater flexibility than conventional phenolic resins in terms of molecular design. They do not release by-products during curing reactions and there is no solvent needed in the resin production. In addition, the bi-functional benzoxazine monomers in particular offer greater flexibility in molecular design than the mono-functional monomers and hence a wide variation in the performance of polybenzoxazines can be achieved.

Ishida and Allen (1996a) studied the properties of bi-functional polybenzoxazine thermosetting based on the ring-opening polymerization of benzoxazine precursors based on aniline and methylamine, namely BA-a and BA-m. Both polybenzoxazines can be synthesized from inexpensive raw materials and polymerized by a ring opening addition reaction with no reaction by-product release. Benzoxazine resin was cured without the aid of the strong acid catalysts normally required by phenolic materials. Dynamic mechanical analysis revealed high moduli and high glass transition temperature with low crosslink densities, i.e. flexural modulus ca. 4.5 GPa, flexural strength ca. 126 MPa, and T_g up to 160°C. These polybenzoxazines possess significantly higher tensile moduli than both phenolics and epoxies yet maintain adequate tensile strength and impact resistance. In addition, low water absorption and good dielectric properties allow these materials to perform well in electronic packaging applications. Ishida (1996) demonstrated that benzoxazine-based resins are expected to yield significant advantages over many other thermosetting materials. The molecular structure of polybenzoxazine allows properties of the cured materials to be controlled for specific requirements of a wide variety of individual applications and the polymer can find applications in areas where polyesters, vinyl esters, epoxies, phenolics, bis-maleimides, cyanate esters, and polyimides are currently used. In addition, a series of benzoxazine resins based on bisphenol-A, paraformaldehyde, and aniline can be synthesized by replacing types of aromatic amines with m-toluidine and 3,5-xylidine (Ishida and Sanders, 2000a; Ishida and Sanders, 2000b; Ishida and Sanders, 2002).

As with any new resin, many parameters must be established before the material may be properly processed. The understanding of thermosetting resins, such as the polybenzoxazines, is complicated by the fact that they are reactive systems with continuously changing properties. Ishida and Rodriguez (1995a) and Ishida and Rodriguez (1995b) studied the curing reactions of benzoxazine precursors based on bisphenol A and aniline (BA-a). This new type of phenolic material cures via a ring-opening mechanism that does not produce any condensation or other reaction by-products. According to differential scanning calorimetry (DSC), the benzoxazine precursor is an autocatalytic-type curing mechanism. The Kissinger and Ozawa models proposed for the curing reaction is valid for the early stages of cure, prior to diffusion control. The activation energy is between 102 kJ/mol and 116 kJ/mol with an overall order of reaction of approximately 2. In addition, the curing of the benzoxazine precursors with catalysts can effectively decrease the induction time for curing and increase the reaction rate. The activation energy values were reported to be 99-107 kJ/mol. The phenol moiety of the ring-opened benzoxazine monomers was reported to have a catalytic effect on the curing reaction, i.e. reducing a reaction induction time and increasing reaction rate. Weak acids such as adipic acid and sebacic acid can also be effectively used as catalysts for benzoxazine resin.

Moreover, curing kinetics of the BA-a resin by dynamic differential scanning calorimetry (DSC) were examined by Jubsilp et al. (2006). The authors reported that the BA-a resin showed only a single curing reaction with the average activation energy of 81–85 kJ/mol, whereas the curing reaction of bisphenol-A and 3,5-xylidine-based benzoxazine resin (BA-35x) was composed of two processes (reactions (1) and (2)), with the average activation energy of 81–87 kJ/mol and 111–113 kJ/mol, respectively. Furthermore, the curing kinetic reaction of the BA-a was found to be the autocatalytic models in nature in good agreement with that reported by Ishida and Rodriguez (1995a) and Ishida and Rodriguez (1995b).

Bi-Functional Polybenzoxazines Based on 2,2′-Dihydroxybiphenyl or 4,4′-Dihydroxybenzophenone and Aniline

The newly developed high-performance polybenzoxazines surpass epoxies, bismaleimides, and can even challenge polyimides in terms of mechanical and thermal properties were investigated by Shen and Ishida (1999). The two benzoxazines were synthesized from 2,2′-dihydroxybiphenyl or 4,4′-dihydroxybenzophenone, paraformaldehyde and aniline (abbreviated as 22P-a and 44O-a, respectively) as shown in Figure 6.

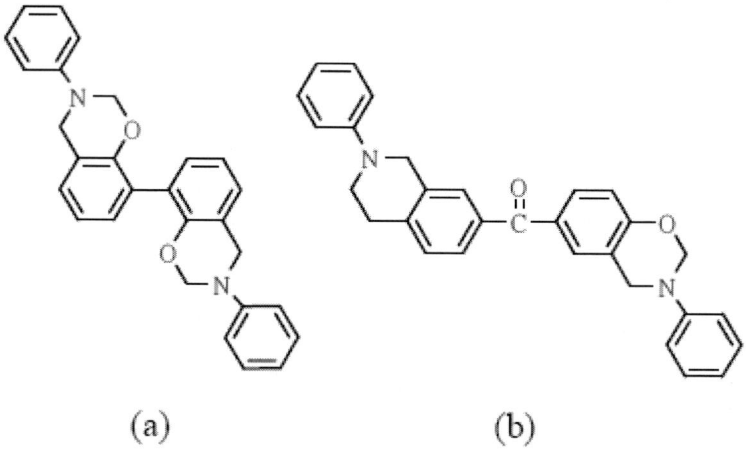

Figure 6. (a) 22P-a and (b) 44O-a-based bi-functional benzoxazine resins.

Table 2. Mechanical and thermal properties of polybenzoxazine (PBZ) compared with general reference of other resins

Materials	Tensile modulus (GPa)	Tensile strength (MPa)	T_g (°C)	T_{d5} (°C)	T_{d10} (°C)	Residual weight (%)
22P-a PBZ (240°C/1h)	-	23	200	250	300	45
44O-a PBZ (290°C/1h)	-	62	340	370	450	65
Other PBZs	4.3-5.2	44-64	150-180	380	-	-
Phenolics	2.7-4.8	35-62	175	-	-	40-50
Epoxies	2.4	28-90	150-261	-	-	40
Bismaleimides	-	-	250-300	450-500	-	50-70
Polyimides	3.2-32.1	30-156	230-380	560-600	-	-

The authors reported that the maximum T_g's achieved in this study for the 22P-a and 44O-a polybenzoxazines are 200°C and 340°C, respectively. Thermal properties of these polybenzoxazines were also studied in terms of isothermal aging and decomposition temperature via thermogravimetric analysis. The optimized curing conditions for the 22P-a and 44O-a polybenzoxazines are 240°C for 1 h and 290°C for 4 h, and the maximum service temperature were found to be 177°C and 290°C, respectively. The thermal properties of the 22P-a and 44O-a benzoxazines compared to other resins are listed in Table 2.

ANILINE-BASED POLYBENZOXAZINE COPOLYMERS

As polymer applications have diversified, the improvement of its processability, thermal and mechanical properties and development of new polymer products, particularly by modification of the existing polymers, continues to be essential. For example, the reactive diluent has been used in copolymer or blend systems by formulating solvent-free applications in coating, adhesion, composites, and electronic encapsulants (Huang and Ishida, 1999; Rimdusit and Ishida, 2000a). In the composite fabrication process, the resin viscosity is an important variable, affecting the flow-out properties and the resin wetting characteristics. Furthermore, diluted resins are employed in some formulations in order to achieve easier handling, increase filler loading and reduce costs. In the past, solvents have been used to lower resin viscosity. However, due to recent stricter environmental regulations on the release of volatile organic compounds (VOC), solvent use has become increasingly restrictive. In recent years, a novel class of thermosetting resin based on a benzoxazine structure has gained substantial attraction because of some of its high thermal and mechanical properties. A relatively low melt viscosity, one of the most useful properties of the benzoxazine resins, results in an ability of the resins to accommodate relatively large quantity of filler while still maintaining their good processability when compared to phenolic resins. Moreover, an excellent property of the benzoxazine resin can blend or alloy with other resins or polymers, i.e. epoxy resin (Ishida and Allen, 1996b; Rimdusit et al., 2011), phenolic resin (Rimdusit et al., 2000b; Jubsilp et al., 2008), urethane prepolyemr (Takeichi et al., 2000; Rimdusit et al., 2005; Rimdusit et al., 2011), polyimide (Takeichi et al., 2005), polycarpolactone (Ishida and Lee, 2001; Zheng et al., 2004), polysiloxane (Ardhyananta et al.,

2008; Ardhyananta et al., 2009), dianhydrides (Jubsilp et al., 2011; Jubsilp et al., 2012; Rimdusit et al., 2012) etc, to obtain high performance polymers.

The synergistic behavior in T_g of polybenzoxazine has been reported in various copolymer and alloy systems as the polybenzoxazine itself possesses relatively low crosslink density comparing with e.g. epoxy having the same type of bisphenol as a starting material. The research in model benzoxazine dimer and trimer structures has indicated that intermolecular and intra-molecular hydrogen bonding is prevalent in this polymer. It is likely that hydrogen bonds in this resin impede the network formation to achieve high crosslink density e.g. by formation of microgel or heterogeneous network. The apparent crosslink density obtained is thus relatively low. The addition of second polymer to form polybenzoxazine copolymer or alloy was often found to enhance crosslink density as reported in polybenzoxazine based on bisphenol-A and aniline, poly(BA-a), As a consequence, its T_g and crosslink density value of poly(BA-a) alloyed with PU elastomer synthesized from polypropylene glycol polyol at a molecular weight of 2000 with toluene diisocyanate (TDI) was significantly enhanced (Rimdusit et al., 2005). Moreover, it was reported that one T_g was observed for the copolymer synthesized from urethane prepolymer and either mono-functional or bi-functional benzoxazine resins. The results suggest high miscibility between the starting polymeric components because of the copolymer reaction.

Copolymers of Ph-a Reactive Diluent and BA-a Benzoxazines

Normally, there are two types of diluents. A reactive diluent is the ingredient that actually undergoes chemical reaction with the resin and becomes part for a formation of the polymeric structure. The other type is a non-reactive diluent which can also lower the viscosity of the base resin but does not take part in the polymer structure formation. The type of resin that has been widely used as a reactive diluent is epoxy resin. This type of reactive diluent was reported to provide good properties and enhanced processing characteristics of the base resin (Huang and Ishida, 1999; Rimdusit and Ishida, 2000a). In addition, urethane prepolymer and unsaturated polyester had also been investigated as reactive diluents of major polymer components (Brzozwski et al., 1997; Takeichi et al., 2000). Recently, a liquid mono-functional benzoxazine resin had been investigated as a reactive diluent of a solid benzoxazine resin. In a report by Huang and Ishida (1999), the viscosity of bisphenol-A-aniline based bi-functional benzoxazine resin (BA-a), a solid benxozazine resin, was reported to be substantially reduced to 20% of that of the neat resin by the use of 25% by mol of 4-cumylphenol and aniline-based mono-functional benzoxazine resin (C-a) as a reactive diluent. Correlations between the observed mechanical properties and network structures of polybenzoxazines as a function of composition were also reported. Glassy state properties, such as stiffness at room temperature, are unaffected by the incorporation of the mono-functional benzoxazine resin. Thermal stability of the mono-functional modified polybenzoxazine below 200°C was also found to be not significantly affected. However, the addition of the C-a resin into the solid BA-a resin was reported to lower a crosslink density of the polymer network and led to a decrease in thermal degradation temperature and residual weight of the polymer hybrids.

Jubsilp et al. (2007) had investigated processability and thermomechanical properties of solid benzoxazine resins based on bisphenol-A and aniline (BA-a) blended with a liquid mono-functional benzoxazine resin based on phenol and aniline (Ph-a) as a reactive diluent. The authors reported that all BA-a:Ph-a resin blends are miscible in nature giving homogenous and transparent liquid mixtures. The viscosity of the BA-a resin can be reduced to one third using only about 10% by weight the Ph-a reactive diluent. The addition of the Ph-a resin into the BA-a resin can also lower the liquefying temperature of the resin mixtures whereas the gel point is marginally decreased. The gel point, which depends on the Ph-a contents and the cure temperature, was determined by the frequency independence of loss tangent in the vicinity of the sol-gel transition. The relaxation exponent values at various the cure temperatures of the blends were found to be 0.24–0.55. In addition, gel time of the BA-a:Ph-a blend systems decreases with increasing temperature according to an Arrhenius relation with activation energy of 60.6-61.5 kJ/mol. Flexural moduli of the BA-a:Ph-a blend systems also increase with the Ph-a content, however, with the sacrifice of their flexural strength and T_g. Furthermore, the kinetic analysis by the dynamic DSC experiments of benzoxazine resin systems such as random co-polybenzoxazine of BA-a type and Ph-a type benzoxazines has also been reported (Sue et al., 2005). The isothermal curing process of the co-polybenzoxazine precursor reveals an autocatalytic-type curing mechanism. The activation energy was found to be 72.11 kJ/mol based on the Kissinger method and 84.06 kJ/mol using the Flynn–Wall–Ozawa method. In the isothermal experiments, the activation energy and Arrhenius pre-exponential based on the Kamal method were reported to be 50.3 kJ/mol and 7959, respectively, whereas the total order of reaction is between 2.66 and 3.03, depending on the isothermal curing temperature.

High Performance Copolymers of Polybenzoxazine-Modified with Dianhydrides

High thermal stability, enhanced char yield, as well as high glass transition temperature (T_g) becomes increasingly important for an advanced composite system such as in an electronic packaging industry. Therefore, the modification of common type polybenzoxazine by blending with other resins or polymers has been studied. As previously reported by Takeichi et al. (2005), aromatic ester (Ar-COOR) can be formed by reaction between hydroxyl group (−OH) of polybenzoxazine based on bisphenol-A and aniline (Poly(BA-a)) and carboxylic acid of poly(amic acid). Recently, Rimdusit and Jubsilp (2007) and Jubsilp et al, (2011) had reported that ester linkage can be formed in dianhydride, i.e. 3,3′,4,4′-benzophenonetetracarboxylic dianhydride (BTDA) blended with the polybenzoxazine based on bisphenol-A and aniline (PBA-a). Fourier transform infrared spectroscopy(FTIR) results confirmed the formation of ester linkage, a structurally flexible functional group, evidenced by the observed peak in the spectrum at 1730 cm^{-1}. Therefore, the bending of poly(BA-a), which is rigid polymeric thermosetting in nature, can be significantly enhanced by an addition of the dianhydride. The single T_g in viscoelastic measurement of this copolymer was also observed, indicating the good compatibility between BTDA and BA-a components. The synergism in T_g of the poly(BA-a) modified with BTDA was reported to be as high as 263°C at the BA-a:BTDA mol ratio of 1:0.67 (60:40 weight ratio). This T_g value was remarkably higher than that of the neat poly(BA-a), i.e. 173°C as seen in Table 3.

Table 3. Thermal properties of poly(BA-a) copolymers modified with BTDA

Copolymers (mol ratio)	T_g [a] (°C)	T_{d5} (°C)	T_{d10} (°C)	Residual weight [b] (%)
BA-a:BTDA (1:0.25)	207	337	381	58
BA-a:BTDA (1:0.33)	221	345	395	60
BA-a:BTDA (1:0.50)	245	363	406	60
BA-a:BTDA (1;0.67)	263	364	414	62
BA-a:BTDA (1:1)	249	352	432	60

[a] T_g from loss modulus curve of dynamic mechanical analysis (DMA).
[b] Residual weight was observed at 800°C.

The improvement of poly(BA-a)'s T_g is due to a substantial increase in crosslink density of the copolymers. Moreover, the degradation temperatures of the copolymers were significant higher than those of the poly(BA-a), i.e. T_{d5} = 334°C and the char yield of the copolymers were in the range of 58% - 62% which implied that the copolymers shows good flammability than the neat poly(BA-a) with the char yield 38% at 800°C.

Recently, effect of aromatic tetracarboxylic dianhydrides, i.e. pyromellitic dianhydride (PMDA), 3,3′,4,4′ biphenyltetracarboxylic dianhydride (s-BPDA), or 3,3′,4,4′ benzophenonetetracarboxylic dianhydride (BTDA) on thermal and mechanical properties of dianhydride-modified bisphenol-A-aniline type polybenzoxazine copolymers were also investigated by Jubsilp et al. (2012). The authors have suggested that the obtained copolymers can be considered as potential candidates for high thermal resistant materials with enhanced mechanical integrity. Glass transition temperature (T_g) of the copolymers connected to rigidity of the dianhydride components was found to be in the order of poly(BA-a):PMDA > poly(BA-a):s-BPDA > poly(BA-a):BTDA. Additionally, the degradation temperature at 10 % weight loss (T_{d10}) of the neat poly(BA-a) was 361°C while the T_{d10} of the copolymers increased in the order of poly(BA-a):PMDA (426°C) > poly(BA-a): s-BPDA (422°C) > poly(BA-a):BTDA (410°C). In other words, the order of thermal stability like degradation temperature with respect to aromatic skeleton of acid dianhydride component is phenylene unit > biphenyl unit > benzophenone unit. Furthermore, the degradation temperature (T_d) value of the neat poly(BA-a) film was observed to significantly increase with an incorporation of aromatic tetracarboxylic dianhydrides which lead to additional crosslinking sites via ester linkage between the anhydride group in the aromatic carboxylic dianhydride and hydroxyl group of the neat poly(BA-a) thus clearly contributed this T_d enhancement.

Rimdusit et al. (2012b) had studied a series of copolymeric films prepared from bisphenol-A-aniline-based bi-functional benzoxazine resin (BA-a) with three biphenyltetracarboxylic dianhydride isomers, i.e., 2,3',3,4'-biphenyl-tetracarboxylic dianhydride (a-BPDA), 2,2',3,3'-biphenyltetracarboxylic dian-hydride (i-BPDA), and 3,3′,4,4′-biphenyltetracarboxylic dianhydride (s-BPDA) through heat curing process. Curing behaviors, dynamic mechanical properties and thermal stability of the obtained polybenzoxazine copolymers were investigated. Their chemical structures analyzed via Fourier transform infrared spectroscopy (FTIR) confirmed ester carbonyl linkage formation between hydroxyl group of the poly(BA-a) and anhydride group in the dianhydride isomer. T_gs of the dianhydride-modified poly(BA-a) increased with poly(BA-a) << i-BPDA < a-BPDA < s-BPDA. T_d of the poly(BA-a) copolymers recorded to be in the range of 365°C-402°C were significantly higher than that of the neat poly(BA-a) i.e. 334°C.

Finally, char yields of the copolymer films were 54%, 55% and 60% for i-BPDA, a-BPDA, and s-BPDA-modified poly(BA-a), respectively, which was much great than that of the neat poly(BA-a), i.e. 38% at 800°C. Therefore, all polybenzoxazine modified with dianhydrides can be considered as good potential candidates for high-temperature resistant materials with outstanding *mechanical integrity* with fire retardant properties.

High Performance Hybrids of Polybenzoxazine-Polyimide

Takeichi et al. (2005) prepared polymeric hybrids using a bi-functional benzoxazine resin based on bisphenol-A and aniline, namely 6,6'-(1-methylethylidene)bis(3,4-dihydro-3-2H-1,3-benzoxazine) (BA-a) and a soluble polyimide (PI) or polyamic acid (PAA, an intermediate compound to synthesize polyimide) derived from bisphenol-A di(phthalic anhydride) ether (BPADA) and oxydianiline (ODA).

The authors reported that an interpenetrating polymer network (IPN) structure was formed in the systems. That is the imidization of PAA occurred *in situ* with the polymerization of BA-a, leading to aromatic ester formation (AR-COOR) that might form between carboxylic acid of PAA and phenolic hydroxyl group of polybenzoxazine during the curing reaction as shown in Figure 7.

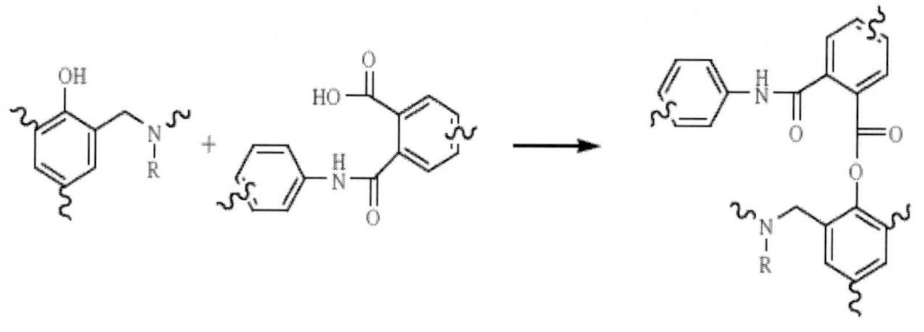

poly(BA-a) PAA

Figure 7. Possible copolymerization between poly(BA-a) and PAA.

In addition, it was suggested that the polymer hybrids from the combination of poly(BA-a) and PAA contain crosslinked structure, which is different from those of poly(BA-a):PI that is supposed to be semi-IPN structures. The polymeric hybrid films from both BA-a:PI and BA-a:PAA had much improved toughness than the neat poly(BA-a). Among the polymeric hybrid films, the films from BA-a:PI at various mass ratios were tougher than those from BA-a:PAA.

The tensile property of the polymer hybrid films was reported to increase with an addition mass fraction of BA-a i.e. 1.39 GPa for PI, 2.02 GPa for PI:BA-a = 50:50, 2.63 GPa for PI:BA-a = 30:70, 3.02 GPa for PI:BA-a = 10:90, and 3.26 GPa for Poly(BA-a) while the films from PAA/BA-a were reported to be too brittle for this measurement. Moreover, the authors found that only one glass transition temperature (T_g) from both loss modulus (E″) and loss tangent (tan δ) was observed in both poly(BA-a):PI and poly(BA-a):PAA hybrid systems, confirming complete miscibility in the blends. The T_gs shift to higher temperature with increasing polyimide content and the T_gs of the polymer hybrids from poly(BA-a):PAA was slightly higher than that of poly(BA-a):PI in the same ratio as listed in Table 4. This is also another evidence indicating the formation of AB cross-linked structure in the case of poly(BA-a):PAA. In addition, the thermal stability, i.e. thermal degradation temperature and char yield of both poly(BA-a):PI and poly(BA-a):PAA hybrid systems were also significantly improved when the polyimide component increased.

Table 4. Viscoelastic properties of the polymer hybrid films from BA-a:PI and BA-a:PAA after cured at 240°C/2 h

Compositions (wt%)	E′ (GPa) at 50°C	T_g (°C) from E″	T_g (°C) from tan δ
BA-a:PI (100:0)	3.18	152	172
BA-a:PI (90:10)	2.17	186	205
BA-a:PI (70:30)	1.17	193	221
BA-a:PI (50:50)	1.10	203	230
BA-a:PI (30:70)	1.09	205	231
BA-a:PI (0:100)	0.59	222	256
BA-a:PAA (70:30)	2.90	196	>210
BA-a:PAA (50:50)	3.55	207	>235
BA-a:PAA (30:70)	3.22	>215	>215

Copolymers from Polybenzoxazine and Epoxy Resin

A copolymer between benzoxazine resin and epoxy resin has also been reported to be a potentially effective measure for thermal and mechanical property enhancement because both polymers have been shown to compatibly undergo hybrid network formation. Some studies had been done to utilize a reactive diluent to lower the liquefying temperature as well as to further reduce the melt viscosity of the as-synthesized benzoxazine resins for processability purposes. Rimdusit and Ishida (2002) investigated the gelation behaviors of low melt viscosity ternary systems of bisphenol-A-aniline benzoxazine (BA-a; B), epoxy (EPON825; E), and phenolic novolac (P) resins. From analysis of Fourier transform mechanical spectroscopy technique (FTMS) in order to study the effect of epoxy diluent on rheological properties of the BEP resin mixtures, the authors found that the liquefying temperature of the uncured resins can be reduced by adding the epoxy resin, besides its major function as a diluent in the molten state of the mixture. The blend renders homogeneous and void-free cured specimens with an outstanding processing characteristic. In addition, low melt viscosity with suitable period of gel time was achieved e.g. viscosity as low as 0.3 Pa s up to 1000 seconds at 100°C. The gel time at 140°C ranges from 5 min to 30 min and less than 5 min at 180°C. The gelation process of these ternary systems shows single activation energy up to their ultimate glass transition temperature signifying single thermal event involved. Furthermore, it can prolong the gel time of the ternary systems without changing their activation energy. Ishida and Allen (1996b) reported that the addition of liquid epoxy, DGEBA, (EPON825) to the polybenzoxazine based on bisphenol and aniline, poly(BA-a), network greatly increased crosslink density of the thermosetting matrix and strongly influenced its mechanical properties. Copolymerization leads to a significant increase in the glass transition temperature (T_g) from about 143°C for poly(BA-a) to about 153°C for an epoxy content of 35% by weight. Further increase in epoxy resin decreases T_g to below that of the neat poly(BA-a). The copolymer also showed improvement in flexural strain at break and flexural strength with only a minimal sacrifice in flexural modulus value. The flexural strength was enhanced from 125 MPa to 170 MPa on increasing the epoxy content to 50% by weight. Kimura et al. (1999) also reported the similar BA-a:epoxy binary systems as investigated by Ishida and Allen (1996b). The curing reaction proceeded without any accelerator. Molding compound showed good thermal stability under 150°C. Above 150°C, the curing reaction proceeded rapidly. The cured epoxy resin showed good heat resistance, water resistance, electrical insulation, and mechanical properties compared to epoxy resin cured by bisphenol-A type novolac.

Furthermore, Rimdusit et al. (2005) showed that the toughness of the alloys of rigid poly(BA-a) and flexible epoxy diluent (EPO732) systematically increased with the amount of the epoxy resin due to the addition of more flexible molecular segments in the polymer hybrids. Although a significant reduction in viscosity and liquefying point was expectedly obtained using the epoxy resin, the resulting benzoxazine-epoxy resin mixtures required higher curing temperatures than that of the neat BA-a resin. The addition of a small fraction of phenolic novolac resin as an initiator into the benzoxazine-epoxy mixtures was reported to be necessary to help lowering their curing temperature (Rimdusit and Ishida, 2000a).

In 2000 Rimdusit and Ishida (2000b) reported synergism in T_g of the ternary systems based on bisphenola-A-aniline benzoxazine (BA-a), epoxy, and phenolic resins. The systems showed the maximum T_g up to about 180°C in mass fraction of BA-a type benzoxazine resin 50%, epoxy resin (EPON825) 40%, and phenolic novolac resin 10%. The addition of a small fraction of phenolic novolac resin enhanced the crosslink density of the fully cured copolymers and the T_g in the copolymers of BA-a and epoxy resins was increased. The molecular rigidity from BA-a resin and the improved crosslink density from epoxy and phenolic contribute to the observed synergistic behavior.

The fully cured copolymers show a T_g as high as 170°C, relatively high degradation temperature, reported at 5% weight loss, i.e. up to 370°C. The char yield of the ternary systems increase significantly compared with the neat epoxy resin. That is due to the fact that both polybenzoxazine and phenolic novolac are known to give higher char yield compared with epoxy resin. The tendency for benzoxazines and phenolic resins to suppress flame propagation makes the ternary copolymer a desirable material in applications where flammability is to be avoided such as interior parts.

Effects of addition of bisphenol-A epoxy resin (EPON828) into various benzoxazine resins based on bisphenol-A and arylamine (aniline, m-toluidine and 3,5-xylidiene), namely BA-a, BA-mt and BA-35x, respectively, have been investigated by Rimdusit et al. (2011). BA-a provided the widest processing window compared with BA-mt, and BA-35x which exhibited the narrowest one. Processing windows of BA-a, BA-mt, and BA-35x were found to be widened with the amount of the epoxy resin. Gel points of benzoxazine-epoxy resin mixtures can be well predicted by an Arrhenius equation, e.g., $t_{gel} = 0.7012\times10^{-7}\exp(10.563/T)$.

Table 5. T_g of arylamine-based polybenzoxazines blended with epoxy resin at various compositions of BA-a/Epoxy, BA-mt/Epoxy, and BA-35x/Epoxy copolymers

Epoxy content (wt%)	T_g (°C)		
	BA-a/Epoxy	BA-mt/Epoxy	BA-35x/epoxy
0	172	210	241
10	178	212	238
20	182	213	235
30	180	209	230
40	174	207	223

From *Dynamic* mechanical analysis (DMA) results, the synergistic behavior with the maximum T_g's was found at the benzoxazine-epoxy composition of 80:20 mass ratio for BA-a/Epoxy and BA-mt/Epoxy copolymer systems as reported in Table 5.

However, it could be noticed that the PBA-a:epoxy copolymer at beyond 30% by weight of epoxy content exhibits a significant decrease in T_g. This behavior is due to the fact that the stoichiometric ratio of components was approached. The unreacted or small molecular weight epoxy molecules might remain in the network formed and might interfere with network formation or acted as a plasticizer in the fully cured network. Furthermore, degradation temperature at 5% weight loss (T_{d5}) of the copolymers were found to increase from 290°C for the neat poly(BA-a) to 350°C for the poly(BA-a):epoxy copolymer with 40% by weight epoxy resin while T_{d5} of the poly(BA-mt):epoxy and the poly(BA-35x):epoxy copolymers were found to be maintained at about 360°C with increasing the amount of epoxy resin. Finally, flexural strength and flexural strain-at-break of all three binary systems were found to increase with increasing amount of the epoxy, meanwhile, the flexural modulus of all copolymers was found to decrease with increasing the amount of the epoxy following the additivity rule.

Moreover, the reaction of epoxy novolac resin cured with benzoxazine resin (BA-a) was investigated under non-isothermal DSC at different heating rates. The study of the curing kinetics contributes to both a better knowledge of process development and an improvement of the quality of final products which is related to the structures of the polymer network. The availability of reliable methods for cure monitoring also plays a crucial role in process control and optimization of the polymer network processing. The kinetic parameters and the kinetic models of the curing processes of the benzoxazine-epoxy novolac mixtures were examined utilizing isoconversional methods,

Flynn-Wall-Ozawa and Friedman methods were utilized by Jubsilp et al. (2010). The benzoxazine-epoxy novolac mixture exhibits two dominant curing processes as evidenced by the presence of two overlapping peaks. The reaction 1 at lower temperature is attributed to the reaction of the benzoxazine monomers, while the reaction 2 corresponds to the formation of side reactions such as etherification reaction of the glycidyl ether by a hydroxyl group of polymerized benzoxazine resin or homopolymerization. The average activation energies of the reaction 1 and reaction 2 were determined to 81 kJ/mol and 118 kJ/mol, respectively. The autocatalytic kinetic model was found to be best description of the investigated curing reactions.

Copolymers from Polybenzoxazine and Polyurethane

Due to high thermal properties, low moisture absorption, better static, and dynamic mechanical properties, as well as a low dielectric constant (DC) of polybenzoxazines, the combination of these materials with polyurethanes (PUs) to enhance their inferior thermal stability was examined. Takeichi et al. (2000) had prepared poly(urethane-benzoxazine) films by blending a benzoxazine monomer (BA-a) and PU that was synthesized from 2,4-tolylene diisocyanate (TDI) and polyethylene adipate polyol (MW ca. 1000) in 2:1 molar ratio. They reported that the main structures in the PU:BA-a coplymer are considered to be urethane formation via a crosslinking between isocyanate group (NCO) of the PU and phenolic hydroxyl group (OH) from ring-opening polymerization of BA-a as well as the allophanate formation via the inter-molecular reaction of the PU. All the films exhibited only one T_g, confirming no phase separation in the obtained poly(urethane-benzoxazine) copolymers due to *in situ* polymerization. The films containing 10% and 15% of BA-a have characteristics of an elastomer, with elongation at break at 244% and 182%, respectively, whereas the films containing more than 20% of BA-a have characteristics of plastics. These elastic films exhibit good resilience with excellent reinstating behavior.

Rimdusit et al. (2005) studied the copolymer between bisphenol-A and aniline type bifunctional benzoxazine resin (BA-a) and urethane prepolymer (PU) prepared from isophorone diisocyanate (IPDI) and polyether polyol with molecular weight of 2000, the synergistic behavior of the T_g of the copolymers was evidently observed. The T_g of the copolymers are noticeably higher than those of the parent polymers, e.g., the copolymer at 30% by weight of PU renders the T_g of approximately 220°C, while those of neat PU and neat poly(BA-a) are reported to be about -70°C and 165°C, respectively. Then, the effect of diisocyanate types, i.e. toluene diisocyanate (TDI), diphenylmethane diisocyanate (MDI) and isophorone diisocyante (IPDI)-based PU on thermal and mechanical properties of poly(BA-a) were also investigated by Rimdusit et al. (2009). The enhancement of the T_g was observed in the copolymers with the PU increased. In addition, the T_gs of the copolymers increased with the weight fraction of polyurethane at least up to 40% by weight of the urethane fraction. This mechanism is likely due to the competition in hydrogen bonding resulting in the crosslinked structure changes. The urethane prepolymer is a softer molecular species having much lower T_g and thus is expected to lower the T_g of the copolymers. Furthermore, the effects of polyol molecular weight e.g. 1000, 2000, 3000 and 5000 on properties of benzoxazine-urethane polymer copolymers have been studied by Rimdusit et al. (2008). They reported that T_g value increased from 160°C in poly(BA-a) to 240°C in poly(BA-a):PU at weight ratio equal to 70:30 as similarly reported by Rimdusit et al. (2005). In addition, the polyol molecular weight shows no effect on the T_g of the BA-a:PU copolymers.

Polybenzoxazine and Polysiloxane Hybrids

Polydimethylsiloxane (PDMS) is one of the most important silicones that have inorganic main chain, which shows good flexibility and oxidative stability, low surface energy, high hydrophobicity, high gas permeability, and good biocompatibility (Abe and Gunji, 2004). Moreover, PDMS can be used even at 450°C as a high-temperature liquid phase for gas chromatography (Takayama and Takeichi, 1994) due to excellent thermal stability. Therefore, the introduction of PDMS into polybenzoxazine matrices is expected to improve the mechanical and thermal properties and the functionalization of the thermosetting polymers as investigated by Takeichi et al. (2001). However, PDMS shows low dispersability in an organic polymer matrix because of its incompatibility. An effective method to enhance the compatibility and dispersability of PDMS is copolymerization. To improve the phase separation of an alloy between poly(BA-a) and polysiloxanes, poly(BA-a)-poly(imide-siloxane) hybrid consisted of hydroxyl group (PISi(OH)) were prepared.

From DMA results, they have confirmed that only one T_g was observed at ca. 300°C caused by an AB-co-cross-linked polymer network is considered to be formed while T_gs of poly(BA-a)-poly(imide-siloxane) hybrid consisted of hydroxyl group (PISi) were observed at ca. 230°C and ca. 300°C, suggesting phase separation between poly(BA-a) and PISi. The tensile properties of poly(BA-a):poly(imide-siloxane) were also examined and they reported that, the elongation at break increased, suggesting the improvement of toughness with increasing the content of poly(imide-siloxane) while the tensile modulus and tensile strength decreased with poly(imide-siloxane) content increased. In addition, the blend with PISi(OH) gives a higher modulus and a lower elongation at the break than the blend with PISi. This is probably because of the difference between the AB-cross-linked polymer and semi-IPN. Therefore, the introduction of poly(imide-siloxane) enhances the toughness of the brittle poly(BA-a), and the effect is more pronounced with PISi, probably because of the formation of the semi-IPN.

In addition, an enhancement of thermal stability, mechanical property, flammability, and flexibility of bisphenol-A-aniline based polybenzoxazine (poly(BA-a)) by adding polydimethylsiloxane (PDMS) has been reported by Ardhyanant et al. (2008). The poly (BA-a)/PDMS hybrids were successfully prepared in situ by the ring-opening polymerization of bisphenol-A-aniline based benzoxazine resin (BA-a) and the sol–gel process of diethoxydi-methylsilane (DEDMS). The authors reported that T_g of the hybrid at the same content of PDMS determined from the maxima of E″ revealed two T_gs, lower T_g (63°C to 72°C) and higher T_g (176°C to 185°C) corresponding to the PDMS and poly(BA-a) components, respectively, suggesting that phase separation at micrometer-scale occurred in the hybrids. The obtained poly(BA-a):PDMS hybrid films with PDMS at 7% and 13% by weight displayed higher degradation temperature at 10% weight loss, i.e. T_{d10} = 352°C and 368°C, respectively, than that of neat poly(BA-a), e.g. T_{d10} = 326°C. Moreover, the char yield or residual weight at 850°C of hybrids showed in the range of 40-46% as evidenced by thermogravimetric analysis, indicating that the flame retardancy is also improved. They additionally reported that the obtained hybrid film render higher tensile strength and elongation at break than neat poly(BA-a) due to the toughening effect of PDMS.

Recently, Ardhyananta et al. (2009) investigated the compatibility and morphology in the blends of poly(BA-a) with polysiloxanes bearing different pendant groups (e.g., methyl and phenyl groups) by synchronizing two reactions; ring-opening polymerization of bisphenol-A and aniline based benzoxazine resins (BA-a) and sol–gel process of diethoxysilanes. The phenyl group of poly (methyl phenyl siloxanes) as PMPS and poly(dimethyl phenyl siloxane) as PDPS enhanced their compatibility with poly(BA-a). The hybrids could improve T_g due to the high crosslinking density of the poly(BA-a) by the plasticizing effect of the polysiloxanes that aid the polymerization of the poly(BA-a). Furthermore, The poly(BA-a):PMPS hybrid displayed the optimum enhancement of the degradation temperature, while the incorporation of PDPS into poly(BA-a) was the most effective for the flame retardancy. Therefore, the thermal properties of the alloys can be controlled by various types of polysiloxanes.

HIGHLY-FILLED ANILINE-BASED POLYBENZOXAZINE COMPOSITES

Reinforcing materials are often incorporated into polymers to modify their properties to meet performance requirements. A wide variation of mechanical, thermal, and physical properties can be developed through an appropriate compounding of polymer and reinforcing materials. For polymeric matrices, phenolic resin was the first synthetic polymer derived from a condensation of phenol and formaldehyde in the presence of acid or alkaline catalysts and up to now is one of the most widely used thermosetting resins (Antony and Pillai, 1994). The major advantages of phenolics are; high temperature resistance, flame retardancy, chemical resistance, dimension stability, and electrical insulation properties. Thus, traditional phenolic resins have been used in many applications such as construction and automotive industries. However, traditional phenolic resins still have many shortcomings. These include the need of strong acid or alkaline catalysts in the synthesis step, the release of by-product such as water or ammonia during the processing step, and their brittleness and limited shelf-life (Jang and Yang, 2000). Therefore, polybenzoxazine or oxazine-based phenolic resins are an alternative to traditional phenolics for the matrix in composite applications due to their excellent mechanical and physical properties. They were investigated as polymer matrices in various composite materials to improve the properties and also to reduce the cost. Interestingly, low melt viscosity is one of the outstanding properties of polybenzoxazine which results in the ability to accommodate relatively large quantity.

Natural Fiber-Filled Polybenzoxazine Composites

High performance composite properties are attributed to the ability of the low viscosity benzoxazine resin (BA-a). Highly filled wood composites made from poly(BA-a) and woodflour with high mechanical properties and reduced water uptake were developed by Jubsilp et al. (2006). The relatively low melt viscosity of the resin enabled the incorporation of a large amount of woodflour up to 75% by weight at its maximum packing. The outstanding compatibility between the woodflour filler and the poly(BA-a) matrix was evidenced by a increase in the glass transition temperature (T_g = 185°C-220°C) when compared with that of neat poly(BA-a), T_g = 160°C. A substantial reduction in water uptake as the amount of the matrix increased due to the inherent low water absorption of the matrix, and substantial interfacial interactions between the woodflour and the matrix. On the other hand, the T_d at 5% weight loss was expectedly observed to decrease with the woodflour content. The char yield of the polybenzoxazine wood showed synergistic characteristics with the peak value of 36%, compared with 28% of the neat resin and 18% of the woodflour. Moreover, a relatively high flexural modulus of 7.3 GPa and flexural strength as high as 75 MPa were achieved in these highly filled system. Furthermore, the utilization of benzoxazine-phenolic novolac alloys as a wood composite matrix was investigated (Rimdusit et al., 2007). The results showed that the incorporation of phenolic novolac into benzoxazine resin (BA-a) can reduce the curing temperature of benzoxazine resin and enhancing the fire resistant characteristics of the resulting copolymers. The limiting oxygen indices (LOIs) of all desirable content of the benzoxazine-phenolic novolac alloys were above the self-extinguishable limit i.e. LOI > 26. In addition, the LOI values were found to moderately increase whereas the rate of burning decreased with increasing the phenolic novolac and high woodflour content at 70% by weight can be incorporated into the copolymer to yield good processing ability and high performance wood composite systems with relatively high flame retardant characteristics.

Recently, properties of experimentally manufactured wood polymer composites based on cashew nut shell liquid (CNSL)-benzoxazine resin (BA-a) copolymer and eastern redcedar woodflour were investigated by Kasemsiri et al. (2011). In this study, The BA-a blended with CNSL at a fixed weight ratio of 85:15 (BA-a:CNSL) was used as a composite matrix. For the curing reaction of the composite, the addition of cashew nut shell liquid and eastern redcedar woodflours effectively reduced the curing temperature and activation energy of BA-a resin. The effects of the woodflour content on the thermal and mechanical increased with optimum amount of woodflour to at 75:25 by weight that attributed to good compatibility between the filler and the BA-a:CNSL matrix. Interestingly, the T_gs of wood composites having eastern redcedar woodflour contents ranging from 50% to 70% by weight were observed to be 162°C to 188°C while the T_g of the BA-a/CNSL matrix was 147°C due to the presence of the phenolic structure in lignin fraction of eastern redcedar particles and the abundance of hydroxyl moieties in the woodflour are believed to provide a composite system with strong interfacial bonding to the polybenzoxazine. Moreover, the wood composites at various contents of particles also showed relatively low percentages of water absorption and thickness swelling. The obtained wood composites based on eastern redcedar woodflour also provided an economic incentive to convert a costly land management problem into the value-added products of wood composite panels.

Long Fiber-Reinforced Polybenzoxazine Composites

In the prepreging process such as a hot melt method, low and fairly constant viscosity is required as the low melt viscosity provides better resin wetting prepregs as well as void-free final product while constant viscosity renders enough time for prepregs processing. The highly processable ternary systems consisted of benzoxazine based on bisphenol-A and aniline (B), epoxy (E), and phenolic novolac (P) resins for carbon fiber composite fabrication was developed by Rimdusit et al. (2009). Phenolic novolac fraction in the ternary mixtures was used to effectively lower the curing temperature and control the processing window of the resin mixtures. The mixtures were able to maintain their processing ability even after a prolonged storing time at room temperature (30°C) up to 270 days is obtained with minimal effect on their processability. The optimal composition of the ternary mixtures for processing of carbon fiber composites consisted of B:E:P at 3:6:2 weight ratio to yield mixture of very low melt viscosity with ultimate T_g and thermal stability in the fully cured polymer. The T_d at 5% weight loss of the BEP362 copolymer is approximately 378°C, comparing with 334°C of the poly(BA-a), Another interesting feature, the char yield of the copolymer at 800°C was found to systematically increase with increasing the phenolic novolac content. This is due to the fact that the phenolic novolac resin possesses the highest char yield (55%) among the three resins used while both benzoxazine resin and epoxy resin render lower char yield i.e., 30%–35% and 15%, respectively. Furthermore, flexural properties of the BEP362 copolymer was observed to be about 135 MPa which higher value compared with that of the poly (BA-a) reported to be 117 MPa while the epoxy possesses the strength of about 110–145 MPa depending on the type of the curing agent. For carbon fiber composite system, the carbon fiber (60% by volume)-reinforced BEP362 copolymer also showed very high enhancement in its modulus value, i.e. storage modulus at 30°C equals to 57.5 GPa, while that of the neat BEP362 copolymer equals to 4.0 GPa. Moreover, TDI-based urethane prepolymer-modified poly(BA-a) reinforced with carbon fiber has been prepared by Rimdusit et al. (2009). They found that the flexural strength of the carbon fiber-reinforced poly(BA-a):PU = 60:40 weight ratio at 80% by weight of the fiber in cross-ply orientation provided relatively high values of about 490 MPa. The flexural modulus slightly decreased from 51 GPa for poly(BA-a) to 48 GPa in the BA-a:PU = 60:40 weight ratio matrix.

In addition, carbon fiber-reinforced benzoxazine composite with high thermal resistance was studied by Shen and Ishida, (1996). Two newly synthesized polybenzoxazines based on 2,2′-dihydroxybiphenyl, paraformaldehyde and aniline (abbreviated 22Pa) and based on 4,4′-dihydroxybenzophenone, paraformaldehyde and aniline (abbreviated 44O-a) were used as a matrix of the composites. The composite with 60% by volume fiber had low void content and gave a high residual weight of more than 87%. Good adhesion between fiber and matrix resulted in a good balance of strength and toughness. The composite showed good mechanical properties, comparable to polyimide (PI) and bismaleimide (BMI) composites. Moreover, PMDA-modified polybenzoxazine (PBA-a) composites reinforced with carbon fiber (CF) have been investigated by Panyawanitchakun et al., (2012) and Rimdusit et al., (2012a). High performance PMDA-modified poly(BA-a) composites reinforced with 65wt% carbon fiber (CF) showed significant enhancement in their thermomechanical properties compared to those of 65wt% CF-reinforced neat poly(BA-a) composite. Storage modulus (E′ at 35°C) and glass transition temperature (T_g) values of all 65wt% CF/PBA-a:PMDA (E′ at 35°C ~ 45 GPa, Tg = 237°C) were remarkably higher than that of the 65wt% CF/PBA-a composite (E′ at 35°C ~ 41 GPa, T_g = 183°C). In addition, the 65wt% CF-reinforced PMDA-modified poly(BA-a) composites passed the highest V-0 classification of UL-94 vertical flame test using the samples as thin as 1 mm. The LOI value of 65wt% CF/poly(BA-a) composite was about 26 and the value was found to be substantially enhanced by an incorporation of PMDA (0.25-1 mol) i.e. 32.5-49.5. This indicates that incorporating PMDA provides a significant effect on promoting the flame resistance of the composite from the unmodified polybenzoxazine. This is due to the superior char forming behavior of the PMDA-modified polybenzoxazine to that of the neat polybenzoxazine. The char formed during combustion can act as a protective layer, which prevents oxygen diffusion to the sample surface thus more oxygen is needed for combustion. Therefore, the obtained composites exhibited great potential applications that required thermomechanical properties and highly efficient self-extinguishing characteristics.

More importantly, carbon fiber prepregs based benzoxazine resin, namely PB1000, have been developed for aircraft interiors by Gurit Company which is a global leader in the development and manufacture of advanced composite materials. They claim that the prepregs or the pre-impregnated composite materials show an excellent mechanical performance, long shelf, shop life and a highest fire protection according to International Federal Acquisition Regulation (FAR)/Justice Acquisition Regulation (JAR). Furthermore, the aerospace group of Henkel Corporation has established Henkel benzoxazine resin like Epsilon 99100 with high mechanical properties and durability, as an effective alternative to the traditional epoxy resin that has used for the aerospace for decades.

In recent years, the development of armor materials has been focused on reducing the weight of the existing armor materials because the reduction of weight could help in saving energy as well as increasing mobility. A light weight ballistic composites from Kevlar^{TM} fiber having poly(BA-a):urethane prepolymer (PU) copolymers as a matrix were investigated by Rimdusit et al, (2011a). The effect of copolymer compositions on the ballistic composite properties was determined. The increase of the elastomeric PU content in the poly(BA-a):PU copolymers resulted in the samples with tougher characteristics; the storage modulus of the Kevlar^{TM}-reinforced poly(BA-a):PU composites increased with increasing the mass fraction of poly(BA-a). A ballistic impact test was also performed on the Kevlar^{TM}-reinforced poly(BA-a):PU composites using a 9 mm handgun and it was found that the optimal composition of poly(BA-a):PU copolymers should be approximately 20% by weight of PU since the results of low level ballistic impact test of the 20-ply Kevlar^{TM}-reinforced composites revealed that only the composite based on the poly(BA-a):PU = 80:20 weight ratio matrix exhibited 100% ballistic penetration resistance.

The suitable thickness having 30 piles of the Kevlar^{TM} cloth can protect the ballistic impact at level II-A, whereas the 50 ply-thick composite was able to protect the ballistic impact of NIJ level III-A. Furthermore, the arrangement of composite panels was also found to significantly affect the ballistic performance of our composites with the thicker and stiffer panel should be placed in the front face of the composite panel assembly to yield best ballistic resistance.

Inorganic Filler-Filled Polybenzoxaizne Composites

Inorganic nanoparticle-filled polymer composites have been classified as materials with the potential to significantly improve many properties over conventional microparticle-filled composites because the filler-matrix interface in these composites might constitute a much greater area and enhance the composite properties to a much greater extent at rather low filler content than microfillers. The effects of nanofiller contents varied from 1 to 10% by weight on the rheological, thermal and mechanical properties of polybenzoxazine/fumed silica nanocomposites have been determined (Rimdusit et al. 2011b). They found that differential scanning calorimetry (DSC) thermograms revealed that fumed silica loading had no significant effect on the thermal curing reaction of benzoxazine resin (BA-a). The rheograms show the shear thinning behaviors of the melted nanocomposite compound and the complex viscosity of the nanocomposites suggested that the liquefying temperature of these molding compounds increased with increasing nano-SiO_2 contents, while the gel-point temperature was not changed even though the amount of nano-SiO_2 was increased. Dynamic mechanical analysis (DMA) revealed that the storage modulus of neat poly(BA-a), i.e. 4.8 GPa was enhanced as the presence only a small amount of nano-SiO_2, i.e. from 5.2 GPa for 1wt% nano-SiO_2 composite to 7.4 GPa for 10wt% nano-SiO_2 composite. The T_g of neat poly(BA-a), T_g = 157°C, was increased with increasing the amount of nano-SiO_2, e.g. T_g = 167°C for 10wt% nano-SiO_2 composite due to the presence of the silica that made the composite stiffer. Additionally, when the nano-SiO_2 was applied, the surface hardness of the neat poly(BA-a), i.e. 368 MPa, was raised from 375 MPa for 1wt% nano-SiO_2 composite to 481 MPa for 10wt% nano-SiO_2 composite. Hence, the addition of nano-SiO_2 to the polybenzoxazine was found to enhance the resistance of the polybenzoxazine deformation.

The effect of microwave processing and thermal curing on the properties of silicon carbide whisker (SiC_w)-filled poly(BA-a) systems has been investigated using an industrial microwave apparatus (Rimdusit et al., 2006). The benzoxazine resin (BA-a) can be cured thermally above 150°C but was hardly cured by microwave irradiation even using a power up to 1 kW. The presence of only 4% by weight of the SiC_w significantly reduces the processing time of the composites from 2 hours at 200°C using the traditional thermal cure to less than 20 minutes using a microwave cure power of 270 W and the mechanical and thermal behaviors of the SiC_w-filled poly(BA-a) show no significant difference with either curing methods. Moreover, the effect of microwave radiation on mass ratio of BA-a resin (B):epoxy resin (E):phenolic novolac resin (P) composites, i.e., BEP361, BEP541, and BEP721, with 10% by weight of SiC_w filler on their thermomechanical properties was also studied (Jubsilp et al., 2009). They have reported that the storage modulus of the BEP composites cured by microwave radiation at 200 W for 45 min was found to be about 6.0 GPa while that of BEP matrices was reported to about 5.0 GPa. In addition, the T_gs of the microwave cured composites obtained from loss modulus peak are slightly lower compared with those of the composites cured by thermal heating due to the different heating mechanisms, i.e. by conduction throughout the whole sample in a thermally cured specimen and a sporadic heating from the surface of tiny SiC_w as well as the inside-out heating mechanism of the microwave during the processing.

CONCLUSION

Polybenzoxazines, a new class of phenolic resins, are widely investigated utilization for potential materials. They can be obtained by the ring-opening polymerization, and possess various advantageous characteristics such properties which are combinations of thermal and flame retardance properties of traditional phenolic resins along with good mechanical performance. Recent studies on the development of polybenzoxazine as a novel type of phenolic resin were reported.

The designing of novel monomers are a promising approach to prepare any type of aniline-based polybenzoxazines taking advantage of the wide selection of mono- and biphenols. Another approach of copolymer and composites is also promising to design polybenzoxazine-based materials with the required performance. The basic studies and applications into various industries are developed as expected.

ACKNOWLEDGMENTS

At many phases of the work described here, the research receives financial supports from the research grants from New Researcher's Grant (2010-2012), Thailand Research Fund and Commission on Higher Education (Grant number: MRG5380077) and by Matching Fund of SWU (Grant number: 083/2554) as well as Research Grants of Faculty of Engineering 2011, SWU

(Grant number: 216/2554) and Graduate School 2011, SWU (Grant number: 217/2554). Additional funding is also provided Higher Education Research Promotion and National Research University Project of Thailand (AM1076A) and from the Rachadaphiseksomphot Endowment Fund Part of the "Strengthen CU's Researcher's Project."

REFERENCES

Abe, Y.; Gunji, T. Oligo- And Polysiloxanes. *Prog. Polym. Sci.*, 2004, 29, 149–182.

Ardhyananta, H.; Wahid, M. H.; Sasak, M.; Agag, T.; Kawauchi, T.; Ismail, H.; Takeichi, T. Performance Enhancement of Polybenzoxazine By Hybridization With Polysiloxane. *Polymer*, 2008, 49, 4585–4591.

Ardhyananta, H.; Kawauchi, T.; Ismail, H.; Takeichi, T. Effect of Pendant Group Of Polysiloxanes On The Thermal And Mechanical Properties Of Polybenzoxazine Hybrids. *Polymer,* 2009, 50, 5959-5969.

Polyurethane MDI Handbook, BASF Corporation, February 2007.

Brzozowski, Z. K.; Szymanska, E; Bratychak, M. M. New Epoxy-Unsaturated Polyester Resin Copolymers. *React. Funct. Polym.*, 1997, 33, 217-224.j.

Holly, F. W.; Cope, A. C. Condensation Products Of Aldehydes And Ketones With o-Aminobenzyl Alcohol And o-Hydroxybenzylamine. *J. Am. Chem. Soc.*, 1944, 66, 1875–1879.

Huang, M. T.; Ishida, H. Dynamic Mechanical Analysis Of Reactive Diluent Modified Benzoxazine-Based Phenolic Resin. *Polym. Polym. Comp.*, 1999, 7, 233-247.

Huang, J.; Zhang, J.; Wang, Fan.; Huang, F. The Curing Reactions Of Ethynyl-Functional Benzoxazine. *React. Funct. Polym.*,2006,66,1395-1403.

Ishida, H. Process For Preparation Of *Benzoxazine* Compounds In Solventless Systems. *US Patent* 5543516, 1996.

Ishida, H. Versatile Phenolic Stands Up To The Heat. *Mod. Plast. Int.*, 1998, 6, 87-90.

Ishida, H. Overview and Historical Background of Polybenzoxazine Research. In: Ishida H., Agag T. (Eds), *Handbook of Benzoxazine Resins*, New York; Elsevier; 2011; Pp 3-81.

Ishida, H.; Allen, D. J. Mechanical Characterization Of Copolymers Based On Benzoxazine And Epoxy. *Polymer,* 1996b, 37, 4487-4495.

Ishida, H.; Allen, D. J. Physical And Mechanical Characterization Of Near-Zero Shrinkage Polybenzoxazine. *J. Polym. Sci. Polym. Phys.*, 1996a, 34, 1019-1030.

Ishida, H.; Lee, Y. –H. Synergism Observed In Polybenzoxazine And Poly(ε-Caprolactone) Blends By Dynamic Mechanical And Thermogravimetric Analysis. *Polymer*, 2001, 42, 6971-6979.

Ishida, H.; Rodriguez, Y. Catalyzing The Curing Reaction Of a New Benzoxazine-Based Phenolic Resin. *J. Appl. Polym. Sci.*, 1995, 58, 1751-1760.

Ishida, H.; Rodriguez, Y. Curing Kinetics Of a New Benzoxazine-Based Phenolic Resin By Differential Scanning Calorimetry. *Polymer,* 1995, 36, 3151-3158.

Ishida, H.; Sander, D. P. Activated Arylamine-Based Polybenzoxazines. *US Patent* 6160079, 2000a.

Ishida, H.; Sanders, D. P. Regioselectivity And Network Structure Of Difunctional Alkyl-Substituted Aromatic Amine-Based Polybenzoxazines. *Macromolecules*, 2000b, 33, 8149-8157.

Ishida, H.; Sanders, D. P. Improved Thermal And Mechanical Properties Of Polybenzoxazines Based On Alkyl-Substituted Aromatic Amines. *J. Polym. Sci. Polym. Phys.*, 2002, 8, 3289-3301.

Jubsilp, C.; Punson, K.; Takeichi, T.; Rimdusit, S. Curing Kinetics Of Benzoxazine Epoxy–Copolymer Investigated By Non–Isothermal Differential Scanning Calorimetry. *Polym. Degrad. Stabili.,* 2010, 95, 918-924.

Jubsilp, C.; Ramsiri, B.; Rimdusit, S. Effects Of Aromatic Carboxylic Dianhydrides On Thermalmechanical Properties Of Polybenzoxazine-Dianhydride Copolymers. *Polym. Eng. Sci.*, 2012, 52, 1640-1648.

Jubsilp, C.; Takeichi, T.; Hiziroglu, S.; Rimdusit, S. Effect Of Resin Compositions On Microwave Processing And Thermophysical Properties Of Benzoxazine-Epoxy-Phenolic Ternary Systems Filled With Silicon Carbide (SiC) Whisker. *Polym. Eng. Sci.*, 2009, 49, 1022–1029.

Jubsilp, C.; Takeichi, T.; Hiziroglu, S.; Rimdusit, S. High Performance Wood Composites Based On Benzoxazine-Epoxy Alloys. *Biores. Technol.*, 2008, 99, 8880–8886.

Jubsilp, C.; Takeichi, T.; Rimdusit, S. Effect Of Novel Benzoxazine Reactive Diluent On Processability And Thermomechanical Characteristics Of Bi-Functional Polybenzoxazine. *J. Appl. Polym. Sci.*, 2007, 104, 2928–2938.

Jubsilp, C.; Takeichi, T.; Rimdusit, S. Property Enhancement Of Polybenzoxazine Modified With Dianhydride. *Polym. Degrad. Stabili.*, 2011, 96, 1047-1053.

Kasemsiri, P.; Hiziroglu, S.; Rimdusit, S. Properties Of Wood Polymer Composites From Eastern Redcedar Particles Reinforced With Benzoxazine Resin/Cashew Nut Shell Liquid Copolymer. *Composites A*, 2011, 42, 1454–1462.

Kimura, H.; Matsumoto, A.; Hasegawa, K,; Fukuda, A. New Thermosetting Resin From Bisphenol A-Based Benzoxazine And Bisoxazoline. *J. Appl. Polym. Sci.*, 1999, 72, 1551-1558.

Ning, X.; Ishida, H. Novel Approach To Phenolic Materials: Synthesis And Characterization Of Phenolic Resins Via Ring Opening Polymerization. *J. Polym. Sci. Chem. Ed.,* 1994a, 32, 1121–1129.

Ning, X.; Ishida, H. Phenolic Materials Via Ring-Opening Polymerization Of Benzoxazines: Effect Of Molecular Structure On Mechanical And Dynamic Mechanical Properties. *J. Polym. Sci. Pol. Phys.*, 1994b, 32, 921-927.

Panyawanitchakun, C.; Jubsilp, C.; Rimdusit, C. Characterization Of High Performance PMDA-Modified Polybenzoxazine Composites Reinforced With Carbon Fiber. *Proceeding of 6th Pure and Applied Chemistry International Conference (PACCON 2012),* Chiang Mai, Thailand, 11-13 January 2012, pp. 1068-1071.

Rimdusit, S.; Bangsen, W.; Kasemsiri, P. Chemorheology And Thermomechanical Characteristics of Benzoxazine-Urethane Copolymers. *J. Appl. Polym. Sci.*, 2011, 121, 3669-3678.

Rimdusit, S.; Ishida, H. Synergism And Multiple Mechanical Relaxations Observed In Ternary Systems Bbased On Benzoxazine, Epoxy, And Phenolic Resins. *J. Polym. Sci. Polym. Phys.*, 2000a, 38, 1687-1698.

Rimdusit, S.; Ishida, H. Development Of New Class Of Electronic Packaging Material Based On Ternary Systems Of Benzoxazine, Epoxy, And Phenolic Resins. *Polymer,* 2000b, 41, 7941-7949.

Rimdusit, S.; Ishida, H. Gelation Study Of High Processability And High Reliability Ternary Systems Based On Benzoxazine, Epoxy, And Phenolic Resins For An Application As Electronic Packaging Materials. *Rheol. Acta.*, 2002, 41, 1-9.

Rimdusit, S.; Jongvisuttisun, P.; Jubsilp, C.; Tanthapanichakoon, W. Highly Processable Ternary Systems Based On Benzoxazine, Epoxy, And Phenolic Resins For Carbon Fiber Composite Processing. *J. Appl. Polym. Sci.*, 2009, 111, 1225–1234.

Rimdusit, S.; Jubsilp, C. Polymer From Anhydride Modified Polybenzoxazine. Thailand Patent pending. Issue Number 5543516, 2007.

Rimdusit, S.; Kampangsaeree, N.; Tanthapanichakoon, W.; Takeichi, T.; Suppakarn, N. Development Of Wood-Substituted Composites From Highly Filled Polybenzoxazine–Phenolic Novolac Alloys. *Polym. Eng. Sci.*, 2007, 47, 140–149.

Rimdusit, S.; Kunopast, P.; Dueramae, I. Thermomechanical Properties Of Arylamine-Based Benzoxazine Resins Alloyed With Epoxy Resin. *Polym. Eng .Sci.,* 2011, 51, 1797-1807.

Rimdusit, S.; Liengvachiranol, C.; Tiptipakorn, S.; Jubsilp, C. Thermomechanical Characteristics Of Benzoxazine–Urethane Copolymers And Their Carbon Fiber–Reinforced Composites. *J. Appl. Polym. Sci.*, 2009, 113, 3823–3830.

Rimdusit, S.; Panyawanitchakun, C.; Jubsilp, C. High Performance PMDA-Modified Polybenzoxazine Composites Reinforced with Carbon Fiber. *Proceeding of MAMIP2012 Asian International Conference on Materials*, Minerals and Polymer, Penange, Malaysia, 23-24 March 2012a, pp. 81-88.

Rimdusit, S.; Pirstpindvong, S.; Tanthapanichakoon, W.; Damrongsakkul, S. Toughening Of Polybenzoxazine By Alloying With Urethane Prepolymer And Flexible Epoxy: A Comparative Study. *Polym. Eng. Sci.,* 2005, 45, 288-296.

Rimdusit, S.; Ramsiri, B.; Jubsilp, C.; Dueramae, I. Characterizations Of Polybenzoxazine Modified With Isomeric Biphenyltetracarboxylic Dianhydrides. *Express Polym. Lett.*, 2012b, 6, 773-782.

Rimdusit, S,; Pathomsap, S.; Kasemsiri, P.; Jubsilp, C.; Tiptipakorn, S. KevlarTM Fiber-Reinforced Polybenzoxazine Alloys For Ballistic Impact Applications. *Engineering Journal*, 2011a, 15, 23-29.

Rimdusit, S.; Punson, K.; Dueramae, I.; Somwangthanaroj, A.; Tiptipakorn, S. Rheological And Thermomechanical Characterizations Of Fumed Silica-Filled Polybenzoxazine Nanocomposites. *Engineering Journal*, 2011b, 15, 27-38.

Rimdusit, S.; Tanthapanichakoon, W.; Jubsilp, C. High Performance Wood Composites From Highly Filled Polybenzoxazine. *J. Appl. Polym. Sci.*, 99, 2006, 1240–1253.

Rimdusit, S.; Jiraprawatthagool, V.; Tiptipakorn, S.; Covavisaruch, S. Characterization *Of SiC Whisker-Filled Polybenzoxazine Cured By Microwave Radiation And Heat. Int. J. Polym. Anal. Charact.*, 2006, 11, 441–453.

Rao, B. S.; Palanisamy, A. Monofunctional Benzoxazine From Cardanol For Bio-Composite Applications. *React. Funct. Polym.*, 2011, 71, 148-154.

Shen, S. B.; Ishida, H. Development And Characterization Of High-Performance Polybenzoxazine Composites. *Polym. Compos.*, 1996, 17, 710-719.

Shen, S. B.; Ishida, H. Dynamic Mechanical And Thermal Characterization Of High-Performance Polybenzoxazines. J. Polym. Sci. Polym. Phys., 1999, 37, 3257–3268.

Su, Y. C.; Yei, D. R.; Chang, F. C. The Kinetics Of B-a and P-a Type Copolybenzoxazine Via The Ring Opening Process. *J. Appl. Polym. Sci.*, 2005, 95, 730-737.

Takayama, Y.; Takeichi, T. Preparation Of Deactivated Metal Capillary For Gas Chromatography. *J. Chromatogr. A*, 1994, 685, 61–78.

Takeichi, T.; Agag, T.; Zeidam, R. Preparation And Properties Of Polybenzoxazine/ Poly(imidesiloxane) Alloys: In Situ Ring-Opening Polymerization Of Benzoxazine In The Presence Of Soluble Poly(Imide-Siloxane)s, *J. Poym. Sci. Pol. Chem.*, 2001, 39, 2633–2641.

Takeichi, T.; Guo, T.; Agag, T. Synthesis And Characterization Of Poly(Urethane-Benzoxazine) Films As Novel Type Of Polyurethane/ Phenolic Resin Composites. *J. Polym. Sci. Polym. Chem.*, 2000, 38, 4165-4176.

Takeichi, T.; Nakamura, K.; Agag, T.; Muto, H. Synthesis Of Cresol-Based Benzoxazine Monomers Containing Allyl Groups And The Properties Of The Polymers Therefrom. *Designed Monom. Polym.*, 2004, 7, 727-740.

Takeichi, T.; Guo, Y.; Rimdusit, S. Performance Improvement Of Polybenzoxazine By Alloying With Polyimide: Effect Of Preparation Method On The Properties. *Polymer*, 2005, 46, 4909-4916.

Wang, Y. X.; Ishida, H. Development Of Low-Viscosity Benzoxazine Resins And Their Polymers. *J. Appl. Polym. Sci.*, 2002, 86, 2953-2966.

Zheng, S.; Lu, H.; Guo, Q. Thermosetting Blends Of Polybenzoxazine And Poly(ε-Caprolactone): Phase Behavior And Intermolecular Specific Interactions. *Macromol. Chem. Phys.*, 2004, 205, 1547–1558.

In: Aniline
Editors: K. Hernandez, M. Holloway
ISBN: 978-1-62417-417-9

Chapter 3

ANILINE PRODUCTION: FROM MULTIPLE STEPS TO ONE STEP

Liangfang Zhu and Changwei Hu[*]
Key Laboratory of Green Chemistry and Technology,
Ministry of Education, College of Chemistry,
Sichuan University, Chengdu, P. R. China

ABSTRACT

The C–N bond-forming reaction, by introducing nitrogen atoms into organic molecules, has an ongoing presence in research at the forefront of chemistry and interrelated fields. Among numerous investigations, the introduction of amino group (-NH_2) into inert aromatic C–H bond to produce aniline is utmost attractive and remains a significant challenge. In this chapter, the examples that concern the synthetic strategies towards aniline production from classical multiple-step method to modern one-step method are reviewed. The classical method for aniline production generally concerns nucleophilic substitution or reductive amination of pre-activated aromatics, but the modern one-step method deals with the direct transfer of -NH_2 group from an amino source, in particular ammonia (NH_3), to aromatic C–H bond. By shortening the synthetic steps, the direct C–H amination process becomes more simple, waste-minimal, and energy-conservative. In addition, insertion of -NH_2 group

[*] Tel: +86 28 85411105; Fax: +86 28 85411105. E-mail address: changweihu@scu.edu.cn or chwehu@mail.sc.cninfo.net.

from simple inorganic aminating agent to aromatic C–H bond gives clue to the formation of organic amines that may be a crucial step to biosynthesis. The influences of several reaction parameters, including source of aminating agent, catalyst, oxidant, solvent, reaction temperature and pressure, are discussed. As a conclusion, the direct synthetic approach of one-step C–H amination is offering a new synthetic strategy towards the clean and effective production of aniline.

Keywords: Aniline, C–N bond-forming, amino transfer, one-step, C–H amination

1. Introduction

Arylamines are abundant in biologically active natural products and medicinally important compounds as well as in materials with useful electrical and mechanical properties. The synthesis of arylamines continues to attract attention by synthetic chemists because the C–N bond-forming reaction has been known to contribute to the ideality of biosynthesis in nature. In the past two decades, significant progress has been achieved within this intriguing research field, either on laboratory or manufacturing scales. [1-8]

The formation of C–N bond involves the introduction of nitrogen atoms into organic molecules and is generally achieved by three synthetic strategies: (I) reductive amination of nitro-compounds, aldehydes, or ketones, (II) transfer of an amino group (-NH_2) from one molecule to another by substitution reaction, and (III) one-step C–H amination *via* simultaneous activation of hydrocarbons and amino source. The former two strategies are indirect process because the C–H bond of hydrocarbons must be pre-activated by reactive functional groups (*i.e.*, -NO_2, -OH, halogens, or transition-metal chlorides) before substitution or reductive amination. The conceptual directness of strategy III endows the one-step C–H amination an attractive and challenging topic in synthetic chemistry from the standpoint of "Green Chemistry". [9, 10] The synthesis of aniline, one of the most valuable chemical intermediates for production of dye, resin, medicine, *etc*., gives good example to illustrate the three synthetic methods.

Commercially, the production of aniline has multiple steps: benzene is nitrated to nitrobenzene by nitric acid, followed by selective reduction or catalytic hydrogenation to aniline by reductive reagent. This multi-step process has many disadvantages, for example, it is energy and time consuming with harsh operation conditions, low atomic efficiency, large amount of by-products, and serious environmental pollutions. Aniline(s) can also be synthesized by reacting Ar-X (X= Cl, Br, OH, etc.) with nucleophiles as diverse as primary and secondary amines, amides, azides, anilines, ammonia and ammonia salts.[6] This method also suffers from serious environmental problems which contradict the "greening" trends of global chemical manufacturing. [10] There is thus a tremendous amount of interest in exploiting new synthetic routes to approach the goal of simple, clean, and energy-conservative production of aniline(s).

More recently, a few of efforts are devoted to the direct synthesis of aniline by one-step amination of benzene. Various aminating agents, including both organic and inorganic amines, have been applied in the direct C–H amination under specific reaction conditions. The direct transfer of $-NH_2$ group from an aminating agent to benzene ring shortens the synthetic step from multiple steps to one step. Most importantly, the direct C–H amination generates minimal wastes, paving the way to a cheaper and improved atom-economical reaction for aniline production.

This chapter aims at summarizing the examples that concern the synthetic strategies towards aniline production from classical multi-step method *via* nucleophilic substitution and reductive amination to modern one-step method *via* direct C–H amination, in particular the direct transfer of $-NH_2$ group from ammonia to benzene ring. The influences of several reaction parameters, including source of aminating agent, catalyst, oxidant, solvent, reaction temperature and pressure, are discussed. It is shown that the direct synthetic approach of one-step C–H amination is offering a new synthetic strategy towards the clean and effective production of aniline. It is hopeful that the perspective taken herein will allow direct C–H amination reaction to be more readily incorporated into synthetic and industrial planning.

2. Multi-Step Amination

2.1. Reductive Amination of Nitrobenzene

The method of reducing nitrobenzene to aniline with iron powder was developed in 1857, which may be the oldest method for commercial aniline production. [11] From the starting material of benzene, multiple reaction steps are involved. Typically, benzene is first nitrated to nitrobenzene with the mixture of nitric/sulphuric acid, and then nitrobenzene is reduced to aniline by iron powder (Scheme 1). In the nitration step, there's serious corrosivity to the equipment because of the strong acidic and oxidative reaction conditions. After being separated from the reaction mixture, the nitrobenzene is inputted into reduction retort where it is reduced by iron powder in acidic media at about 100 °C. The resulting mixture is neutralized with slaked lime ($Ca(OH)_2$), obtaining crude aniline after separation. The product aniline is finally obtained by repeated washing. The yield of the product is largely determined by the quantity of iron powder. Obviously, this multi-step process has many disadvantages. For example, it has low atomic efficiency; it is energy and time consuming with harsh operation conditions; it costs large amount of iron powder (reactant) and acids with the generation of large amount of by-products such as salts; and it causes serious environmental pollutions and instrumental corrosivity.

This method was soon replaced by continuous catalytic hydrogenation, which had been developed into classic manufacturing process for aniline and other bulk arylamines production (Scheme 2). [12]

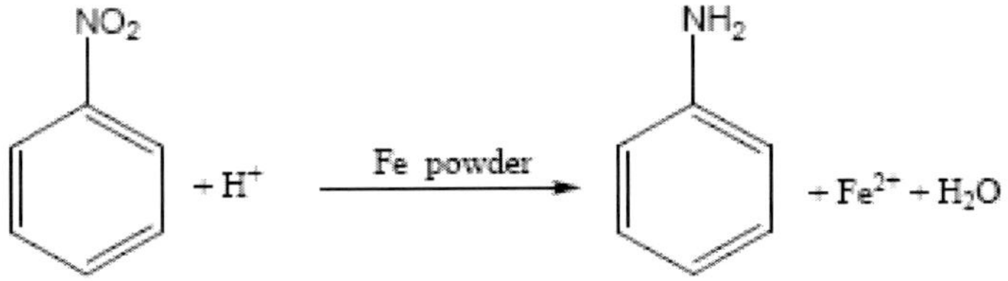

Scheme 1. Reduction of nitrobenzene with Fe powder.

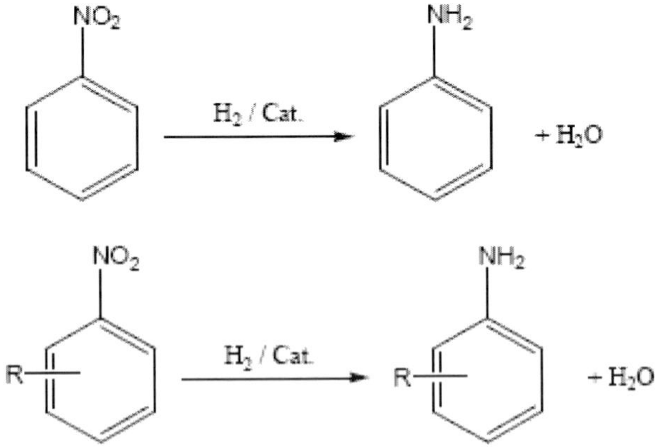

Scheme 2. Catalytic hydrogenation of nitrobenzene to aniline with hydrogen.

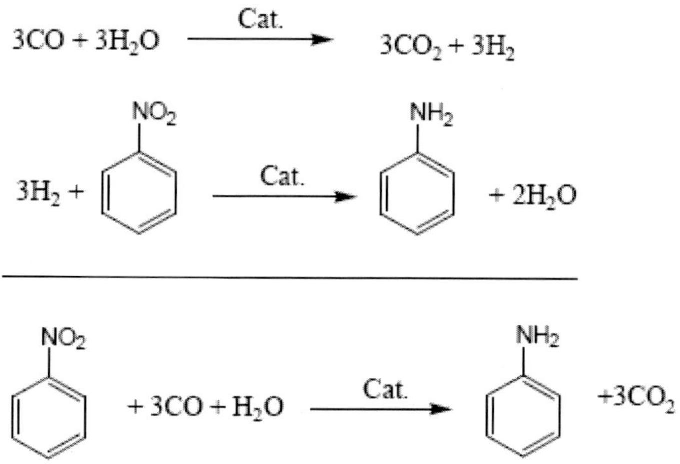

Scheme 3. Catalytic hydrogenation of nitrobenzene to aniline with water gas.

In comparison with the chemical reduction method, the catalytic hydrogenation of nitrobenzene improved the atomic efficiency and avoided the salts pollution to a greater extent. Therefore, it was long practiced for aniline production and so far, a lot of work had been done on improving the performance of catalyst and the reaction craft. [13-29] For example, a great deal of technological considerations was devoted to the selectivity, stability/regenerability, susceptibility to poisons, and mechanical strength of the catalyst. Moreover, the improvement of reaction craft mainly focused on removing the heat from the reaction system because of the high exothermicity of the hydrogenation reaction.

In general, the heterogeneous catalyst used in the catalytic hydrogenation includes two types: i) supported transition metal catalyst (*i.e.*, SiO_2-supported Cu, Co or Ni), ii) supported noble metal catalyst (*i.e.*, Pt, Pd or Rh supported on Al_2O_3, SiO_2 or activated carbon). In recent years, advanced materials such as molecular sieves, [30, 31] carbon nanotube, [32, 33] and reduced graphene oxide [34] were emerged as catalyst or support for preparing the hydrogenation catalyst. To enhance the selectivity to the reductive amination, the catalyst is generally modified with a second metal (inhibitor) by preventing ring hydrogenation. The hydrogenation processing can be carried out in vapor or liquid phases. The vapor-phase processing is employed in either fixed-bed or fluid-bed reactors, where Cu/SiO_2 is typically employed as the catalyst. Instead of hydrogen, natural gas can be used for nitrobenzene hydrogenation. [35] For liquid-phase hydrogenation, the same catalyst as employed in the vapor-phase can be used. [36] Moreover, water gas (CO/H_2O) can also be used as reductive gas for nitrobenzene hydrogenation with carbonyl complexes of Fe, Ru or Rh as catalyst (Scheme 3). [37-39] Although great progress on the catalytic hydrogenation of nitrobenzene have been achieved, the environmental issues and/or feedstock cost concerns limit the development of the multi-step synthesis of aniline in the modern "green" manufacturing.

2.2. Substitution Amination of Phenol

To decrease the cost of raw material in conventional nitrobenzene hydrogenation, the method of catalytic ammonolysis of phenol was developed in 1962 and industrialized in 1970 by Halcon Company (Scheme 4).

The reaction is performed in gas phase under high temperature and high pressure conditions. A series of solid-acid catalysts such as titania-zironia, [40] titania-silica, [41] alumina/modified aluminas, [12, 42-44] and β-, MFI- or ZSM-5-typed zeolites [45-48] have shown activity in the amination, and some of them have applied into practical use. In addition, a lot of work has been done on suppressing the catalyst deactivation and cost for utility. For example, because the strong acid site on the solid-acid catalyst may cause strong adsorption of aniline, [47] the catalyst with relatively weaker acid strength was used to suppress the catalyst deactivation.

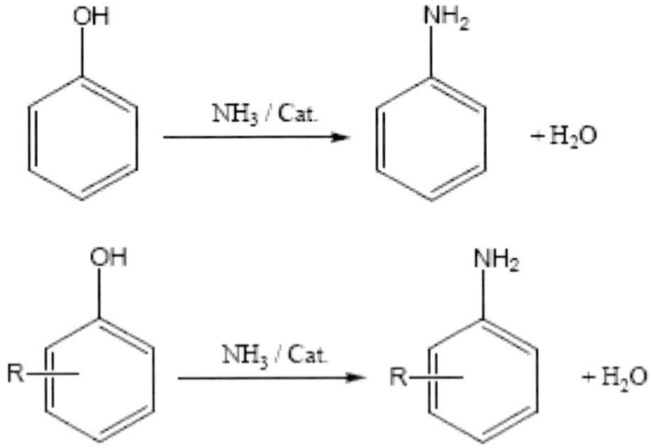

Scheme 4. Synthesis of aniline(s) from phenol(s) *via* substitution of hydroxyl group with ammonia.

2.3. Nucleophilic Substitution of Aryl Halides

An alternative commercial process for aniline(s) production is based on the substitution of halogen group (*i.e.*, -Cl) in aryl halides with ammonia at high temperature (300~600 °C) and high ammonia pressure (Scheme 5). [49, 50]

Some heterogeneous catalysts such as alumino-slicate or Cu-containing zeolite (mordenite and L) have been applied in practical use. In this approach, however, the halogen group is a leaving group, which decreases the atomic efficiency and increases the wastes of the synthetic approach, especially in industrial scales. Moreover, the aryl halides, used herein as raw materials, have to be pre-synthesized through multiple steps that are known to be environmentally polluted processes.

Recently, Xia and co-workers [51] modified the synthetic method by using bromine and metal oxides as mediators, which is suitable for laboratory synthesis of aniline. That is, benzene was firstly brominated to bromobenzene catalyzed with Fe powder or ZrO_2, and bromobenzene was then ammonolyzed over CuO/TiO_2 cataloreactant. This approach gave out overall aniline yield of ca. 90% with selectivity of ca. 95% under a relatively lower temperature (e.g. below 200 °C).

Although some environmental issues exist, the nucleophilic aromatic substitution of aryl halides is still a long established reaction in anilines synthesis.

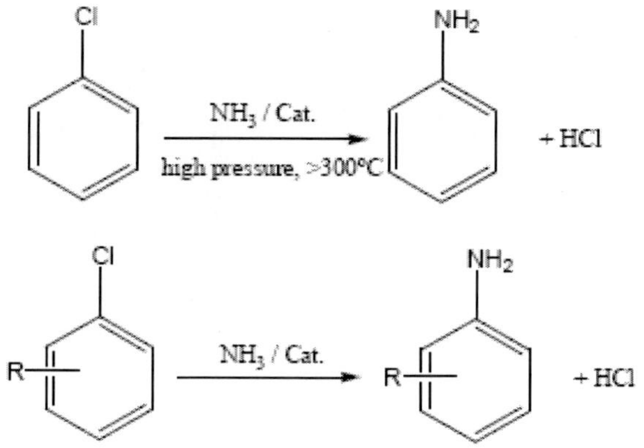

Scheme 5. Synthesis of aniline(s) from aryl halide(s) *via* substitution of halogen group with ammonia.

Transition-metal catalyzed (especially palladium-catalyzed) C–N cross-coupling reaction between Ar-X (X = Cl, Br, *etc.*) and nucleophiles has emerged as a powerful synthetic tool for the synthesis of anilines. [6] A wide range of nucleophiles such as primary and secondary amines, [52-61] amides, [62-65] azides, [66-69] hydrazine, [70] anilines, [71-75] and ammonia salts [76] were well documented over the last two decades. Despite the increasing complexity of the amino source molecules that can be fashioned using these methods, the use of the conceptually simplest N-based nucleophiles — ammonia —has remained a significant challenge. Especially in liquid-phase ammonolysis of Ar-X, [77-84] the facile binding of ammonia with metals further complicates the amination process. [85]

2.4. Other Amination

Aniline can also be synthesized by rearrangement of N-methyl aniline over HZSM-5 and HY, [86] or amination of cyclohexanol and/or cyclohexanone followed by non-oxidative dehydrogenation over Ni-based zeolites [87] or oxidative dehydrogenation over mixed oxide catalysts (V-Mo/Al_2O_3 or Ni-Mo/Al_2O_3) [88, 89] (Scheme 6). The latter two cases suffer from low selectivity towards aniline because of the existence of side-reaction such as dehydration to form cyclohexene. Furthermore, high reaction temperature (> 350 °C) is always involved in the two-step amination process, limiting the yield of product thermodynamically.

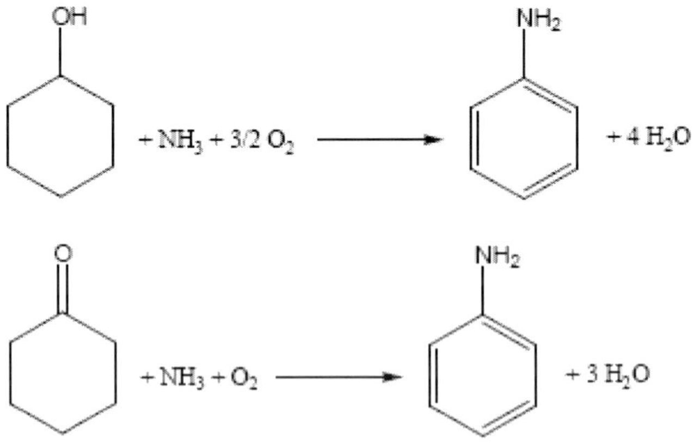

Scheme 6. Amination of cyclohexanol and cyclohexanol to aniline by oxidative dehydrogenation.

3. ONE-STEP C–H AMINATION *VIA* DIRECT AMINO TRANSFER

The direct formation of C–N bond by reaction between a nitrogen atom and an un-activated C–H bond is a highly atom-economical process. [90-92] Generally speaking, this one-step process involves the activation of C–H bond of an organic molecule and insertion of an amino group from an amino source. Some early but successful results on direct C–H amination have been summarized in some leading reviews. [93-95] In the context of the synthesis of aniline(s), however, it requires finding suitable conditions for the generation of highly active species to induce the activation of inert aromatic C–H bond because of its high energy. Actually, while adequate oxygenases for selective C–H hydroxylation do exist in nature, the analogous direct nitrogen transfer has not been found in any biosynthetic pathway so far. [96] Thus, it remains a real challenge for the direct transfer of an amino group from an aminating agent onto an un-activated benzene ring. Along with the screening of effective catalyst for the direct C–H amination, a range of amino source, containing both organic amines and inorganic derivatives from ammonia, are investigated.

3.1. Organic Amines as Aminating Agent

Organic amines, in the forms of RNH_2 or RR'NH, have been used as the amino source for the direct transfer of nitrogen atom to aromatic ring. In Diversi's work, [97] diethylamine was used as the amino source for the direct amination of benzene to N, N-diethylaniline with the formation of many by-products such as N,N,N',N'-tetraethylurea, pentamethyl-cyclopentadience, metallic mercury and an unidentified iridium compound. (Scheme 7) The aromatic C–H bond of benzene and N–H bond of diethylamine are simultaneously activated by the mixture of Ir(I)/Hg(II)-based system at 130 °C under N_2 atmosphere.

The phenylmercury(II) species, formed from the electrophilic attack of benzene by $HgSO_4$, are attributed to the subsequent amination. However, the unreacted Ir(I)/Hg(II) reagents are remained in the reaction mixture which makes the separation of the product difficult. Moreover, the involved radical mechanism entangles the product distribution although no halogenated materials or intermediates are involved in.

$$C_6H_6 + NHEt_2 \xrightarrow[130\ ^{o}C,\ N_2,\ 24\ h]{[Ir(\eta^5\text{-}C_5Me_5)(CO)_2],\ HgSO_4} C_6H_5NEt_2$$
$$+ Et_2NCONEt_2 + C_5Me_5H + Hg + [Hg] + [Ir]$$

Scheme 7. Direct amination of benzene with diethylamine in the presence of an Ir(I)/Hg(II) based system. [97].

Stern et al. [98] reported the method of direct formation of aromatic C–N bond by aminating nitrobenzene with aniline, promoted by tetramethylammonium hydroxide dehydrate (TMA(H)·$2H_2O$) at 50 °C under anaerobic conditions. Marquet and coworkers [99] described the hydrogen nucleophilic aromatic photosubstitution of *m*-dinitrobenzene with primary amines (aniline or *n*-butylamine) or amides (acetamide or benzamide) with the help of fluoride anion (TBAF, tetrabutyl ammonium fluoride). In these two cases, the aromatic C–H bond is activated by the promoters of TMA(H)·$2H_2O$ or TBAF in spite of the complexity of molecules involved.

3.2. Ammonia as Aminating Agent

Ammonia (NH_3) is the simplest source of amino group and is also a bulk raw material of chemistry. The low cost and wide availability of ammonia make it an extremely attractive source of amino group for synthesis. [100] Direct amination of benzene with ammonia is practically attractive from the viewpoints of both atom economy and the direct useness of raw material. However, the strength of the N–H bonds in ammonia (107 kcal mol-1) makes the "N–H activation" challenging. [101] There is thus tremendous amount of interest in employing catalyst system and reaction craft for simultaneous activation of N–H bond of ammonia and C–H bond of benzene, aiming at the target of one-step synthesis of aniline from benzene. Several noble-metal and transition-metal catalysts have been found to exhibit activity for simultaneously activating the N–H bond of gaseous ammonia and C–H bond of benzene, wherein rigorous reaction conditions (e.g., high T and/or P) are required.

3.2.1. Dehydrogenation Amination

As early as 1917, for the first time Wibaut [102] already proposed the direct amination of benzene with gaseous ammonia over a nickel/iron catalyst over the temperature range of 550~600 °C under atmospheric pressure (Scheme 8).

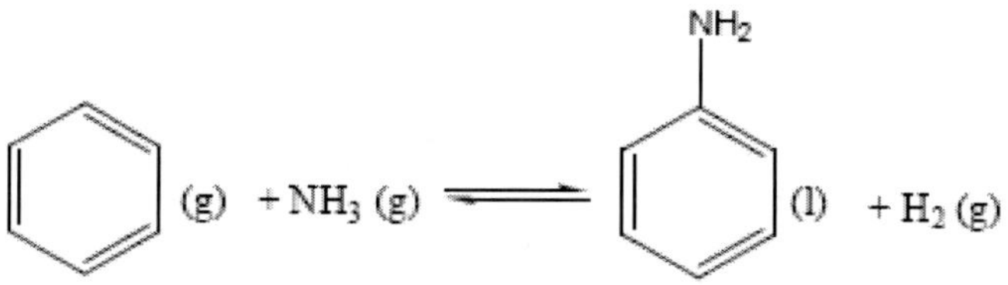

Scheme 8. Direct amination of benzene to aniline with gaseous ammonia.

The atomic efficiency of the reaction is as high as 98%, with formation of the only by-product of hydrogen. However, the conversion of benzene is rather low due to the equilibrium limit, which limits the academic and industrial developments of the dehydrogenation amination.

3.2.2. Lattice Oxygen as Oxidant

By selective removal of hydrogen from the reaction between benzene and ammonia, the equilibrium limit can be shifted and improved benzene conversion is enabled. The only by-product of this reaction — water — is one of the most benign chemical to the environment.

Since the sixties of the last century there have been several reports on the reaction of benzene, ammonia, and a kind of oxidative species to directly synthesize aniline. At the earliest, lattice oxygen was used as the oxidant for selective removal of hydrogen (Scheme 9).

In 1958, CA patent 553988 [103] disclosed a process for preparing aniline by oxy-amination of benzene with gaseous ammonia in the presence of a reducible metal oxide (oxides of Fe, Co, Ni, Sn, Sb, Bi, or Cu) at a temperature range of 100 to 1000 °C. In 1960, US patent 2948755 [104] described the direct synthesis of aniline from benzene and ammonia at 400~500 °C under 1~200 bar in the presence of a bi-metallic catalyst (especially ammonium molybdate and cupric oxide).

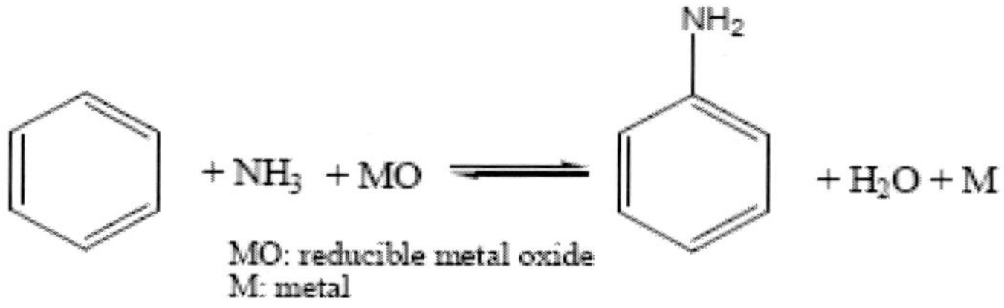

Scheme 9. Oxy-amination of benzene to aniline with ammonia using lattice oxygen as the oxidant.

In 1994, Hara [105] described the oxy-amination of benzene with ammonia over Group VIIIA metals containing catalyst. For example, the reaction over Fe/SiO_2 catalyst at 400 °C gave out an aniline yield of 0.85% with a selectivity of 97.3%.

The important milestone for the development of oxy-amination reaction may be the discovery of the adjustable Ni/NiO catalyst system. In 1975, US patent 3919155 and 3929889 [106] disclosed the use of Ni/NiO at 150~500 °C under 10~1000 bar for the oxy-amination of benzene to aniline. During the amination, NiO is partially reduced to elemental Ni at reducing atmosphere (e.g., H_2), while elemental Ni formed by this process is then re-oxidized partially to nickel oxide in an oxidizing atmosphere (e.g., oxygen, air or water). In 1977, Del Pesco et al. (DuPont Company) [37] described a similar process in US patent 4031106 using $Ni/NiO/ZrO_2$ as the cataloreactant. By using the redox Ni/NiO-based catalyst system, the hydrogen produced from reaction of benzene and ammonia is sacrificed, hence increasing the conversion of benzene. The best yield of aniline is 13% after reacting benzene and ammonia at 300 °C under 30 bar for 4 h in a batch reactor. In 2000, Poojary et al. [107] prepared a catalyst by mixing a noble metal selected from Pd, Rh, It, and/or Ru with a reducible metal oxide. A best benzene conversion of at least 5% with an aniline selectivity of more than 90% was obtained. The catalyst can be regenerated by re-oxidation with oxygen without a substantial loss of performance. Likewise, Hagemeyer and Desrosiers et al. [108, 109] reported a catalyst system where the combination of a noble metal (Rh, Ir, Pd, or Ru) and a reducible metal oxide (Ni or Co oxide) gave out a stable benzene conversion of 10% and a selectivity of above 95% after reaction at 300 °C under 300 bar for 2 h (*i.e.* over $Rh/Ni\text{-}Mn/K\text{-}TiO_2$ catalyst).

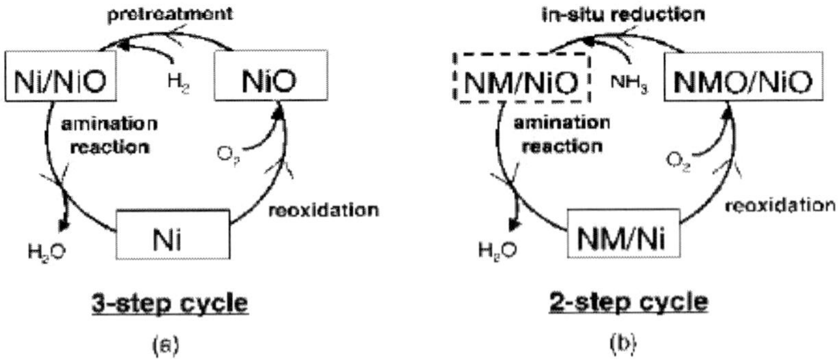

Scheme 10. Possible mechanism for the oxy-amination of benzene with ammonia over an adjustable Ni/NiO catalytic system. [109]. Copyright © 2002, Elsevier.

Correspondingly, the authors proposed a mechanism for the catalytic cycle of Ni/NiO-based catalyst in the oxy-amination (Scheme 10). Some literatures also gave explanation of the oxy-amination process under the role of the NiO/ZrO_2 catalyst system. [110, 111] However, although aniline can be prepared from benzene and gaseous ammonia in a single step with the assistance of lattice oxygen, the high pressure presents a formidable barrier to commercialization due to the high energy cost.

3.2.3. Molecular Oxygen as Oxidant

Introduction of molecular oxygen (O_2) into the amination system also promotes the conversion of benzene by shifting the reaction equilibrium (Scheme 11). The atomic efficiency of this reaction is about 84%, with formation of the only-product — water.

In Thomas' work, [103] oxidative-amination of benzene was performed in the presence of NH_3 and O_2 over a platinum catalyst at about 1000 °C. In Becker and Hölderich's work [112, 113], NH_3 and O_2 were co-feeded in a plug flow reactor or in a continuously stirred tank reactor where supported Group VIII metals (*i.e.*, Pd, Pt, Rh, Ph, or Ni supported on Al_2O_3; or CuO supported on ZrO_2) were used as the catalyst. The amination occurred at elevated temperature and pressure. For example, a benzene conversion of 0.98% and a selectivity to aniline of 99% were obtained over Pd/NaX-zeolite catalyst after reacting at 500 °C under 30 bar for 4 h (n_{C6H6}: n_{NH3} = 1:55). Over Rh/α-Al_2O_3 catalyst, the conversion of benzene was 1.4%, with an aniline selectivity of 100%, after reacting at 500 °C under 50 bar for 6.5 h. In 1999, US patent 5861536 [114] disclosed the use of a supported metal and binucleating complex in the oxy-amination of benzene with O_2 as the terminal oxidant performing at a temperature range from 100 to 450 °C under a pressure above 4 bar. PCT WO 9910311 [115] described a similar process, wherein the catalyst used comprises at least one metal selected from transition metals, lanthanides and actinides, preferably Cu, Pt, V, Rh, and Pd.

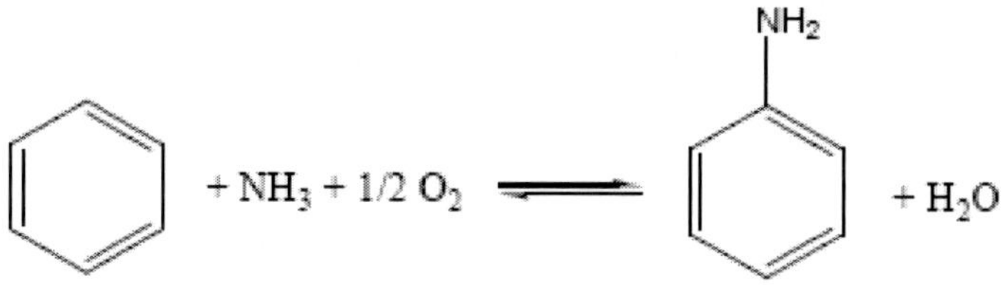

Scheme 11. Oxy-amination of benzene with ammonia using gaseous oxygen as the terminal oxidant.

However, the selectivity of aniline obtained is lower than 85%. In 2000, Stitt et al. in PCT WO 0009473 [116] used vanadium oxide-based catalyst for the oxy-amination of benzene, giving out a selectivity of aniline of 96% (450 ^{o}C, 2~9 bar).

3.2.4. Hydrogen Peroxide as Oxidant

Obviously, when gaseous ammonia is used as the amino source in the oxidative amination, lattice oxygen or gaseous oxygen is generally used as the terminal oxidants. However, the reaction conditions are rather harsh because high temperature (*i.e.*, 100~1000 °C) and/or high pressure (*i.e.*, 10~1000 bar) are required. In particular, noble metals are always involved for substantially activate the N–H bond of ammonia and C–H bond of benzene. The high cost of the catalyst material also limits the commercialization of the high T/P oxy-amination process.

More seriously, only low yield and/or selectivity of aniline are obtained. Therefore, there remains significant challenge for employing new synthetic process for "greener" synthesis of aniline considering the energy and raw materials costs.

Aqueous ammonia ($NH_3{\cdot}H_2O$) is also a good candidate as amino source for the amination reaction. Using $NH_3{\cdot}H_2O$ as the aminating agent, the amination can be performed in liquid-phase under mild conditions (e.g., low temperature, atmospheric pressure). For example, Yoshida and Yuzawa [117] reported the direct amination of benzene by aqeuous ammonia on Pt/TiO_2 photocatalyst, although the yield of aniline was still relatively low (0.028%). In 2003, Hu group [118, 119] proposed a new synthetic approach for preparing aniline directly from benzene and $NH_3{\cdot}H_2O$ using hydrogen peroxide (H_2O_2, 25~28 *wt%*) as the terminal oxidant (Scheme 12). One of the problematic issues in this synthetic route is that there are competitive reactions within the reaction system: oxy-amination of benzene to aniline and hydroxylation of benzene to phenol.

In Chen et al.'s work, aniline formed dominantly (relatively to hydroxylation of benzene to phenol) after reacting benzene, $NH_3{\cdot}H_2O$, and H_2O_2 over Ni-Zr-Ce/Al_2O_3 catalyst at 50 ^{o}C under atmospheric pressure for 2 h. [119] Moreover, increasing the feeding ratio ($n_{benzene}$: $n_{NH3{\cdot}H2O}$) from 1:1 to 1:5 increased the yield of aniline significantly without much effect on the formation of phenol. In comparison with the gas-phase oxy-amination, the liquid-phase amination of benzene occurs under a milder reaction condition, providing a new "greener" chemical way for the single-step production of aniline.

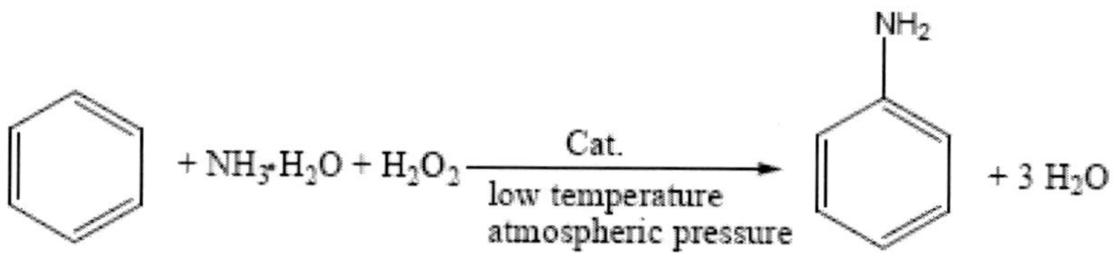

Scheme 12. Oxy-amination of benzene to aniline with aqueous ammonia using hydrogen peroxide as the terminal oxidant.

Obviously, the catalyst employed in the liquid-phase oxy-amination of benzene should be more active because it should be capable to simultaneously activate the C–H bond of benzene, N–H bond of $NH_3 \cdot H_2O$, and O–O bond of H_2O_2. Some multi-component catalysts such as Zr-Ni/Al_2O_3, [120] V-Ni/Al_2O_3, [121] Mo-Ni/Al_2O_3, and Mn-Ni/Al_2O_3 [118] have shown activity. More recently, metal-modified TS-1 catalysts (metal: Cu, Ni, Ce, V, Co, or Fe) have been applied in the liquid oxy-amination of benzene to aniline. [122] The contribution of catalytic species on the oxy-amination was extensively investigated: TS-1 is active for activating both C–H bond of benzene and N–H bond of ammonia, while the latter bond activation is reinforced by the presence of framework Ti species. Most importantly, the doping of copper on TS-1 leads to the formation of framework Cu-O-Ti specie on the catalyst, promoting the chemical adsorption of ammonia on the solid catalyst and contributing significantly to the enhanced performance for activating N–H bond of ammonia. As a best result, acceptable yield (1% based on the amount of benzene used) and selectivity (88%) for aniline were obtained on Cu-TS-1 (2.4 wt% of Cu content) after optimizing the reaction conditions (70 °C, atmospheric pressure, 2 h). Although the yield of aniline is still modest, we may expect the recycling of benzene and aqueous ammonia to enhance the utilization of raw materials.

In addition to delicate catalyst design, a catalytic distillation process was performed over the V-Ni/Al_2O_3 catalyst to enhance the yield of aniline (Figure 1). [123] By separating the product aniline from the reactants (benzene and ammonia) and catalyst, the chemical equilibrium was shifted in the desired direction, resulting in a higher conversion of reactants and a greater yield of product. The effects of several parameters (including the packing manners of the catalytic column, the feed ratio, the distillation temperature, and the reaction time on the amination) were well studied, which provides an alternative approach for enhancing the yield and selectivity of aniline for given catalyst.

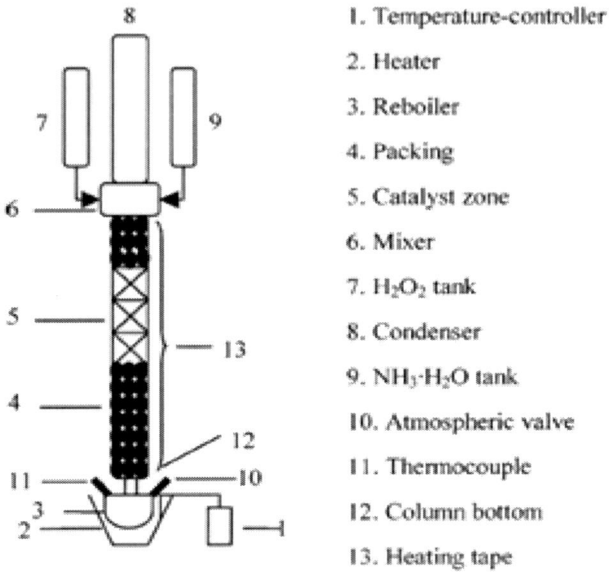

Figure 1. Scheme of the catalytic distillation setup for the oxy-amination of benzene to aniline with $NH_3{\cdot}H_2O$ and H_2O_2.[123].

3.3. Hydroxylamine as Aminating Agent

In 1994, Mantegazza et al. [124] described the direct synthesis of hydroxylamine (NH_2OH) with high yield by oxidation of ammonia with hydrogen peroxide in the presence of Ti-silicalite at low temperature under atmospheric pressure. With suitable selection of solvent, no potential deep oxidation of ammonia occurs, and no byproduct forms. The formed *in situ* NH_2OH showed aminating ability for ammoximation of cyclohexanone to cyclohexane oxime over Ti-containing molecular sieves. [125, 126] Thus, it provided a potential benign supply of hydroxylamine for the amination of benzene (Scheme 13). However, free-hydroxylamine is unstable and its derivatives in the form of salts are more safely handled. Leading investigations on the NH_2OH-involed amination use hydroxylamine sulfate or hydrochloride as the aminating agent.

In 2000, Kuznetsova et al. [127] reported the amination of benzene with hydroxylamine sulfate in the presence of homogeneous transition metal redox catalysts (salts of V(V), Fe(III), or Mo(VI)) in a closed system (90 ^oC).

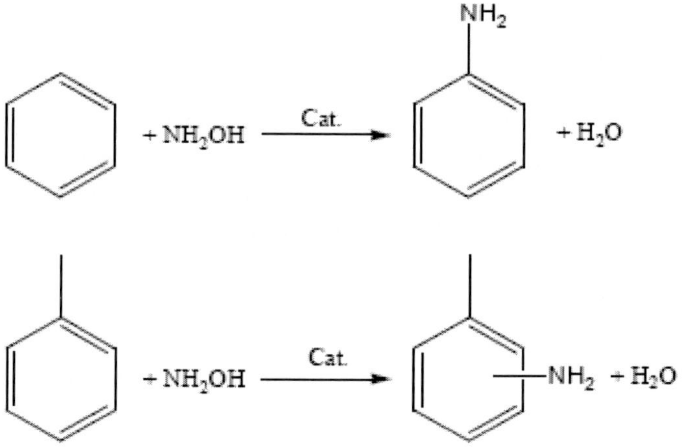

Scheme 13. Amination of benzene or toluene to aniline or toluidines with hydroxylamine.

The reaction was carried out in acetic acid–water or sulfuric acid–acetic acid–water media, obtaining a maximum aniline yield of about 27%. When toluene was used as the substrate, *o*-, *m*-, and *p*-toluidines formed without obvious selectivities. It seems that the $-NH_2$ group in structure of NH_2OH is more readily to be transferred to the aromatic ring. Hu group [128] gave out a detailed explanation on the NH_2OH-involed amination chemistry, where the decomposition of NH_2OH ($NH_2OH{\cdot}HCl$) was found to be a crucial factor affecting the amination process. The effects of reaction temperature, the acetic acid concentration, as well as the amount and the valence of vanadium catalyst on the decomposition of NH_2OH and amination process were systematically studied. The oxy-amination process was performed in acetic acid-water media in open air (80 °C) over a series of homogeneous or heterogeneous vanadium-based catalysts.[128-133] In comparison with the closed system, the yield of aniline (68 mol%) and toluidines (overall 50% for *o*-, *m*-, and *p*-products) increased significantly under optimal reaction conditions. Moreover, by-products (*i.e.*, phenol, biphenyl, chlorobenzene, and some other N-substituted aromatic compounds), as formed in a closed system, were not detected in the air-open system. The acidity of the reaction media in the latter air-open system was much milder than that of the closed amination system. A reaction mechanism about the amination chemistry is proposed.[128, 134, 135] Further work mainly focus on the development of more effective catalyst system to enhance the yield and selectivity of aniline. [136-139] For example, a Mn(II)-dampy-MCM-41 catalyst gave out a best aniline yield of 74.5 mol% with selectivity to aniline of 100%. [138] Some electrochemical methods for direct amination of benzene to aniline have also emerged, whereas, strong acidic media are required as well. [140, 141]

In spite of the high yield of aniline in the NH_2OH-aminating system, the potential risks in handling poisonous and explosive NH_2OH, as well as the high cost of NH_2OH, limit the commercialization of this amination process for producing aniline. Even so, the detailed mechanistic study on the direct amination of benzene with hydroxylamine has helped us discern such amination chemistry and provided information for synthesizing aniline directly from benzene, ammonia, and hydrogen peroxide in the future. It may be the most promising route for the replacement of the traditional aniline production.

CONCLUSION

The development of new environmental favorable routes for the production of commercially relevant chemical intermediates and products is an area of considerable interest. In this limited chapter, the synthetic strategies towards aniline production from classical multi-step method to modern one-step method are reviewed.

The classical multi-step synthesis of aniline requires pre-installing an active functional group on benzene ring, which is subsequently aminated by reducing agent or nucleophilically substituted by amino group. This multi-step process suffers from several environmental, energy, and cost issues, which contradict the "greening" trends of global chemical manufacturing. The direct C–H amination reaction, by directly transferring an amino group from aminating agent to benzene ring, shortens the synthetic route to a single step. In particular, the direct synthesis of aniline from the very starting materials of benzene, ammonia, and hydrogen peroxide, represents a promising direction towards aniline production.

Besides providing the direct methods to achieve these conceptually simple transformations, numerous efforts are devoted to the development of new catalyst system to simultaneously activate the C–H bond of benzene, N–H bond of ammonia, and O–O bond of hydrogen peroxide. These recent developments have provided solutions to one of the longest standing challenges in C–N bond-forming process. Future investigation may be directed to the development of new types of catalyst, as well as the integration of catalyst with the process intensification to further enhance considerably the yield of aniline.

REFERENCES

[1] Downing, R. S.; Kunkeler, P. J.; vanBekkum, H., Catalytic syntheses of aromatic amines. *Catalysis Today* 1997, *37* (2), 121-136.

[2] Ma, D.; Cai, Q., Copper/amino acid catalyzed cross-couplings of aryl and vinyl halides with nucleophiles. *Accounts of Chemical Research* 2008, *41* (11), 1450-1460.

[3] Aubin, Y.; Fischmeister, C.; Thomas, C. M.; Renaud, J.-L., Direct amination of aryl halides with ammonia. *Chemical Society Reviews* 2010, *39* (11), 4130-4145.

[4] Tsuchiya, K., Synthesis of Macrocyclic Aromatic Amines via C-N Coupling Reaction. *Journal of Synthetic Organic Chemistry Japan* 2011, *69* (2), 169-170.

[5] Crozet, D.; Urrutigoity, M.; Kalck, P., Recent advances in amine synthesis by catalytic hydroaminomethylation of alkenes. *ChemCat-Chem* 2011, *3* (7), 1102-1118.

[6] Belfield, A. J.; Brown, G. R.; Foubister, A. J., Recent synthetic advances in the nucleophilic amination of benzenes. *Tetrahedron* 1999, *55* (38), 11399-11428.

[7] Beccalli, E. M.; Broggini, G.; Fasana, A.; Rigamonti, M., Palladium-catalyzed C-N bond formation via direct C-H bond functionalization. Recent developments in heterocyclic synthesis. *Journal of Organo-metallic Chemistry* 2011, *696* (1), 277-295.

[8] Borodkin, C. I.; Shubin, V. G., Nitrenium ions and problem of direct electrophilic amination of aromatic compounds. *Russian Journal of Organic Chemistry* 2005, *41* (4), 473-504.

[9] Anastas, P. T.; Warner, J. C., *Green Chemistry: Theory and Practice*, Oxford University Press: New York, 1998.

[10] Baker, R. T.; Tumas, W., Toward Greener Chemistry. *Science* 1999, *284* (5419), 1477-1479.

[11] Weissermel, K.; Arpe, H.-J., *Industrial Organic Chemistry*, 3rd Edition, John Wiley and Sons, Chap. 13, 1997.

[12] Yasuhara, M.; Matsunaga, F., Preparation of Anilines. *US Patent 4987260* 1991.

[13] Bancir, K. Gas phase prepn. of aniline-by catalytic hydrogenation of nitrobenzene with hydrogen. CS8003665-A, 1983.

[14] High purity aniline prodn. by catalytic hydrogenation of nitrobenzene in presence of alkali metal hydroxide or (bi)carbonate or zinc acetate or nitrate. *JP89036460-B2*; *JP58004750-A*, 1983.

[15] Manufacture of aniline by hydrogenation of mononitrobenzene in a trickle bed reactor, comprises feeding liquid nitrobenzene and hydrogen with or without recycle reactor product through catalytic reactor. *RD508024-A*, 2006.

[16] Gorbunov, B. N.; Utrobin, A. N. Low-temperature method for aniline through hydrogenation of nitrobenzene. *RU2337904-C2*, 2008.

[17] Immel, O.; Waldmann, H.; Braden, R.; Froehlich, C.; Friedhofen, G.; Niemeier, W. Nitrobenzene gas-phase hydrogenation to aniline by hydrogenation of nitrobenzene cpds. using catalysts comprising palladium@ on graphite. *US Patent 5304525-A*, 1994.

[18] Li, J.; Yang, H. Palladium nano material useful for catalyzing hydrogenation of nitrobenzene into aniline, comprises pure palladium snowflake-shaped nano particle with single appearance and even dispersion. *CN101972651-A*, 2011.

[19] Nagata, T.; Watanabe, K.; Kono, Y.; Tamaki, A.; Kobayashi, T. Continuous prepn. of high purity aniline by catalytic hydrogenation of nitrobenzene, using palladium@ (platinum@) catalyst deposited on lipophilic carbon@. *EP476404-A2*, 1992.

[20] Shen, Z.; Gu, L.; Chen, J. Fluidized-bed catalyst for preparing aniline by hydrogenation of nitrobenzene. *CN1657162-A*; *CN1285407-C*, 2006.

[21] Shen, Z.; Xiao, J.; Yao, H. Preparation method of fluidized bed catalyst for aniline production by gas phase hydrogenation of nitrobenzene. *CN1768941-A*; *CN100368085-C*, 2006.

[22] Zhao, F.; Cai, S.; Wang, H.; Meng, X. Method for synthesizing amino benzene by hydrogenation of nitrobenzene in low temperature. *CN1803761-A*, 2006.

[23] Zhou, Z.; Fu, Q.; Chen, D.; Li, D.; Sun, Z.; Li, L. Catalyst for fluidized bed gaseous hydrogenation of nitrobenzene to produce aniline, and process therefor. *CN1861254-A*, 2007.

[24] Didillon, B.; Lepeltier, F.; Candy, J. P.; Boitiaux, J. P.; Basset, J. M., Surface organometallic chemistry on metals-catalytic-hydrogenation of nitrobenzene into aniline on silica supported rhodium modified by tetrabutyltin. In *Progress in Catalysis*, 1992; Vol. 73, pp 23-30.

[25] Yeong, K. K.; Gavriilidis, A.; Zapf, R.; Hessel, V., Experimental studies of nitrobenzene hydrogenation in a microstructured falling film reactor. *Chemical Engineering Science* 2004, *59* (16), 3491-3494.

[26] Vilella, I. M.; de Miguel, S. R.; Scelza, O. A., Study of the performance of Pt catalysts supported on activated carbon felt and granular carbon for

nitrobenzene hydrogenation. *Chemical Engineering Journal* 2005, *114* (1-3), 33-38.

[27] Sun, Z.; Zhao, Y.; Xie, Y.; Tao, R.; Zhang, H.; Huang, C.; Liu, Z., The solvent-free selective hydrogenation of nitrobenzene to aniline: an unexpected catalytic activity of ultrafine Pt nanoparticles deposited on carbon nanotubes. *Green Chemistry* 2010, *12* (6), 1007-1011.

[28] Wang, J.; Yuan, Z.; Nie, R.; Hou, Z.; Zheng, X., Hydrogenation of nitrobenzene to aniline over silica gel supported nickel catalysts. *Industrial and Engineering Chemistry Research* 2010, *49* (10), 4664-4669.

[29] Mohan, V.; Pramod, C. V.; Suresh, M.; Reddy, K. H. P.; Raju, B. D.; Rao, K. S. R., Advantage of Ni/SBA-15 catalyst over Ni/MgO catalyst in terms of catalyst stability due to release of water during nitrobenzene hydrogenation to aniline. *Catalysis Communications* 2012, *18*, 89-92.

[30] Liu, C.; Tan, R.; Yu, N.; Yin, D., Pt-Pd bi-metal nanoparticles captured and stabilized by imine groups in a periodic mesoporous organosilica of SBA-15 for hydrogenation of nitrobenzene. *Microporous and Mesoporous Materials* 2010, *131* (1-3), 162-169.

[31] Liu, H.; Lu, G.; Guo, Y.; Wang, Y.; Guo, Y., Synthesis of spherical-like Pt-MCM-41 meso-materials with high catalytic performance for hydrogenation of nitrobenzene. *Journal of Colloid and Interface Science* 2010, *346* (2), 486-493.

[32] Li, C. H.; Yu, Z. X.; Yao, K. F.; Ji, S. F.; Liang, J., Nitrobenzene hydrogenation with carbon nanotube-supported platinum catalyst under mild conditions. *Journal of Molecular Catalysis A-Chemical* 2005, *226* (1), 101-105.

[33] Zhao, Y.; Li, C.-H.; Yu, Z.-X.; Yao, K.-F.; Ji, S.-F.; Liang, J., Effect of microstructures of Pt catalysts supported on carbon nanotubes (CNTs) and activated carbon (AC) for nitrobenzene hydrogenation. *Materials Chemistry and Physics* 2007, *103* (2-3), 225-229.

[34] Gao, Y.; Ma, D.; Wang, C.; Guan, J.; Bao, X., Reduced graphene oxide as a catalyst for hydrogenation of nitrobenzene at room temperature. *Chemical Communications* 2011, *47* (8), 2432-2434.

[35] Kricsfalussy, Z.; Stammann, G.; Waldmann, H., *Chemie-Ingenieur-Technik* 1994, *66*, 832-834.

[36] Relvas, J.; Andrade, R.; Freire, F. G.; Lemos, F.; Araujo, P.; Pinho, M. J.; Nunes, C. P.; Ribeiro, F. R., Liquid Phase hydrogenation of nitrobenzene over an industrial Ni/SiO_2 supported catalyst. *Catalysis Today* 2008, *133*, 828-835.

[37] DelPesco, T. W.; Del, W., Synthesis of aromatic amines by reaction of aromatic compounds with ammonia. *US Patent 4031106 and 4001260* 1977.

[38] Chaloner, P. A.; Esteruelas, M. A.; Joo, F.; Oro, L. A., *Homogeneous Hydrogenation, Catalysis by Metal Complexes*. Kluwer Academic Publishers: Durdrecht, 1994; Vol. 15.

[39] Ragaini, F.; Cenini, S.; Gasperini, M., Reduction of nitrobenzene to aniline by CO/H_2O, catalysed by $Ru_3(CO)_{12}$/chelating diimines. *Journal of Molecular Catalysis A-Chemical* 2001, *174* (1-2), 51-57.

[40] Kiyoura, T.; Takahashi, T., *JP48067229, 1973, to Mitsui Toatsu Chemicals Inc.* 1973.

[41] Tanabe, K.; Ito, M.; Sato, M., High acidity of silica–titania and its high catalytic activity for amination of phenol with ammonia *J. Chem. Soc., Chem. Commun.* 1973, 676-677.

[42] Becker, M.; Sachs, H. M., Process for the production of variable amounts of DPA and aniline using only phenol and ammonia as the feedstock. *US Patent 4480127 A* 1984.

[43] Yasuhara, M.; Matsunaga, F., *European Patent No. 3212750A2, 1989, to Nitsui Petrochemical Industries Ltd.* 1989.

[44] Schutz, A. A.; Cullo, L. A., *PCT Patent, WO 9305010 A1* 1993.

[45] Chang, C. D.; Lang, W. H., *European Patent No. 62542 A1* 1982.

[46] Katada, N.; Kuroda, S.; Niwa, M., High catalytic activity for synthesis of aniline from phenol and ammonia found on gallium-containing MFI. *Applied Catalysis A-General* 1999, *180* (1-2), L1-L3.

[47] Katada, N.; Iijima, S.; Igi, H.; Niwa, M., Synthesis of aniline from phenol and ammonia over zeolite beta. *Studies in Surface Science and Catalysis* 1997, *105*, 1227-1234.

[48] Miyamoto, T.; Katada, N.; Kim, J.-H.; Niwa, M., Acidic property of MFI-type gallosilicate determined by temperature-programmed desorption of ammonia. *Journal of Physical Chemistry B* 1998, *102*, 6738-6745.

[49] Jones, D. G.; Hill, C., Production of aromatic amines by ammonolysis. *US Patent 3231616*, 1966, to Socony Mobil Oil Company Inc.

[50] Burgers, M. H. W.; Van Bekkum, H., Substitution of halobenzenes to aniline or phenol, catalysed by copper-exchangedzeolites. Selectivity improvements and reaction pathway. *Studies in Surface Science and Catalysis* 1994, *84*, 1981-1988.

[51] Zhai, H. S.; Lin, H. Q.; Lu, X.; Xia, X. H., Highly efficient amination of benzene to aniline mediated by bromine with metal oxide as cataloreactant. *Chemistry Letters* 2006, *35* (12), 1358-1359.

[52] Reddy, N. P.; Tanaka, M., Palladium-catalyzed amination of aryl chlorides. *Tetrahedron Letters* 1997, *38* (27), 4807-4810.

[53] Djakovitch, L.; Wagner, M.; Kohler, K., Amination of aryl bromides catalysed by supported palladium. *Journal of Organometallic Chemistry* 1999, *592* (2), 225-234.

[54] Desmarets, C.; Schneider, R.; Fort, Y., Nickel-catalysed synthesis of 3-chloroanilines and chloro aminopyridines via cross-coupling reactions of aryl and heteroaryl dichlorides with amines. *Tetrahedron Letters* 2001, *42* (2), 247-250.

[55] Desmarets, C.; Schneider, R.; Fort, Y., Nickel-catalysed sequential amination of aryl- and heteroaryl di- and trichlorides. *Tetrahedron* 2001, *57* (36), 7657-7664.

[56] Jiao, J.; Zhang, X. R.; Chang, N. H.; Wang, J.; Wei, J. F.; Shi, X. Y.; Chen, Z. G., A facile and practical copper powder-catalyzed, organic solvent- and ligand-free Ullmann amination of aryl halides. *Journal of Organic Chemistry* 2011, *76* (4), 1180-1183.

[57] Yang, Z.; Chen, Y.; Ni, C. Y.; Ren, Z. G.; Wang, H. F.; Li, H. X.; Lang, J. P., Arylamine-solvated copper(I)/iodide coordination polymers cooperatively formed by the amination of aryl iodides and the assembly of [Cu_2I_2] species with 1,2-bis(4-pyridyl)ethane. *Inorganic Chemistry Communications* 2011, *14* (9), 1537-1540.

[58] Hirai, Y.; Uozumi, Y., Heterogeneous aromatic amination of aryl halides with arylamines in water with PS-PEG resin-supported palladium complexes. *Chemistry-an Asian Journal* 2010, *5* (8), 1788-1795.

[59] Kim, M.; Chang, S., Rhodium(NHC)-catalyzed amination of aryl bromides. *Organic Letters* 2010, *12* (7), 1640-1643.

[60] Desmarets, C.; Schneider, R.; Fort, Y., Nickel-mediated amination chemistry. Part 1: Efficient aminations of (het)aryl 1,3-di and 1,3,5-trichlorides. *Tetrahedron Letters* 2000, *41* (16), 2875-2879.

[61] Manolikakes, G.; Gavryushin, A.; Knochel, P., An efficient silane-promoted nickel-catalyzed amination of aryl and heteroaryl chlorides. *Journal of Organic Chemistry* 2008, *73* (4), 1429-1434.

[62] Tao, C. Z.; Li, J.; Fu, Y.; Liu, L.; Guo, Q. X., Copper-catalyzed synthesis of primary arylamines from aryl halides and 2,2,2-trifluoro-acetamide. *Tetrahedron Letters* 2008, *49* (1), 70-75.

[63] Anjanappa, P.; Mullick, D.; Selvakumar, K.; Sivakumar, M., 2-(Trimethylsilyl)ethanesulfonyl amide as a new ammonia equivalent for palladium-catalyzed amination of aryl halides. *Tetrahedron Letters* 2008, *49* (31), 4585-4587.

[64] Romero, M.; Harrak, Y.; Basset, J.; Orue, J. A.; Pujol, M. D., Direct synthesis of primary arylamines via C-N cross-coupling of aryl bromides and triflates with amides. *Tetrahedron* 2009, *65* (10), 1951-1956.

[65] Mullick, D.; Anjanappa, P.; Selvakumar, K.; Ruckmani, K.; Sivakumar, M., Carbamic acid 2-trimethylsilylethyl ester as a new ammonia equivalent for palladium-catalyzed amination of aryl halides. *Tetrahedron Letters* 2010, *51* (46), 5984-5987.

[66] Zhao, H.; Fu, H.; Qiao, R., Copper-catalyzed direct amination of ortho-functionalized haloarenes with sodium azide as the amino source. *Journal of Organic Chemistry* 2010, *75* (10), 3311-3316.

[67] Messaoudi, S.; Brion, J. D.; Alami, M., Copper-catalyzed direct amination of halo(hetero)arenes with sodium azide as the amino source. *Mini-Reviews in Organic Chemistry* 2011, *8* (4), 448-454.

[68] Goriya, Y.; Ramana, C. V., The [Cu]-catalyzed S(N)AR reactions: direct amination of electron deficient aryl halides with sodium azide and the synthesis of arylthioethers under Cu(II)-ascorbate redox system. *Tetrahedron* 2010, *66* (38), 7642-7650.

[69] Maejima, T.; Shimoda, Y.; Nozaki, K.; Mori, S.; Sawama, Y.; Monguchi, Y.; Sajiki, H., One-pot aromatic amination based on carbonenitrogen coupling reaction between aryl halides and azido compound. *Tetrahedron* 2012, *68*, 1712-1722.

[70] Pastyrikova, T.; Iakobson, G.; Vida, N.; Pohl, R.; Beier, P., Direct amination of nitro(pentafluorosulfanyl)benzenes through vicarious nucleophilic substitution of hydrogen. *European Journal of Organic Chemistry* 2012, (11), 2123-2126.

[71] Yang, S. C.; Hung, C. W., Palladium-catalyzed amination of allylic alcohols using anilines. *Journal of Organic Chemistry* 1999, *64* (14), 5000-5001.

[72] Li, Z. H.; Xia, Z. N.; Chen, G., Catalyst-free amination of 2-chloronicotinic acid in water under microwave irradiation. *Journal of Chemical Research* 2011, (12), 709-711.

[73] Rodriguez, S.; Qu, B.; Haddad, N.; Reeves, D. C.; Tang, W.; Lee, H.; Krishnamurthy, D.; Senanayake, C. H., Oxaphosphole-based mono-phosphorus ligands for palladium-catalyzed amination reactions. *Advanced Synthesis and Catalysis* 2011, *353* (4), 533-537.

[74] Zhang, Y.; Silverman, R. B., Direct Amination of gamma-Halo-beta-ketoesters with Anilines. *Journal of Organic Chemistry* 2012, *77* (7), 3462-3467.

[75] Hoi, K. H.; Calimsiz, S.; Froese, R. D. J.; Hopkinson, A. C.; Organ, M. G., Amination with Pd-NHC complexes: rate and computational studies involving substituted aniline substrates. *Chemistry-A European Journal* 2012, *18* (1), 145-151.

[76] Kim, J.; Chang, S., Ammonium salts as an inexpensive and convenient nitrogen source in the Cu-catalyzed amination of aryl halides at room temperature. *Chemical Communications* 2008, (26), 3052-3054.

[77] Lang, F. R.; Zewge, D.; Houpis, I. N.; Volante, R. P., Amination of aryl halides using copper catalysis. *Tetrahedron Letters* 2001, *42* (19), 3251-3254.

[78] Vaganova, T. A.; Kusov, S. Z.; Rodionov, V. I.; Shundrina, I. K.; Malykhina, E. V., Selective mono- and diamination of polyf luorinated benzenes and pyridines with liquid ammonia. *Russian Chemical Bulletin, International Edition* 2007, *56* (11), 2239-2246.

[79] Q. Shen, Q.; Hartwig, J. F., Palladium-catalyzed coupling of ammonia and lithium amide with aryl halids. *Journal of American Chemistry Society* 2006, *128*, 10028-10029.

[80] Wang, D.; Cai, Q.; Ding, K., An efficient copper-catalyzed amination of aryl halides by aqueous ammonia. *Advanced Synthesis and Catalysis* 2009, *351* (11-12), 1722-1726.

[81] Guo, Z.; Guo, J.; Song, Y.; Wang, L.; Zou, G., Hemilabile-coordinated copper promoted amination of aryl halides with ammonia in aqueous ethylene glycol under atmosphere pressure. *Applied Organometallic Chemistry* 2009, *23* (4), 150-153.

[82] Schulz, T.; Torborg, C.; Enthaler, S.; Schaeffner, B.; Dumrath, A.; Spannenberg, A.; Neumann, H.; Boerner, A.; Beller, M., A general palladium-catalyzed amination of aryl halides with ammonia. *Chemistry-A European Journal* 2009, *15* (18), 4528-4533.

[83] Meng, F.; Zhu, X. H.; Li, Y.; Xie, J. W.; Wang, B.; Yao, J. H.; Wan, Y. Q., Efficient copper-catalyzed direct amination of aryl halides using aqueous ammonia in water. *European Journal Organic Chemistry* 2010, 6149-6152.

[84] Zhu, Y.; Wei, Y., A simple and efficient copper-catalyzed amination of aryl halides by aqueous ammonia in water. *Canadian Journal of Chemistry-Revue Canadienne De Chimie* 2011, *89* (6), 645-649.

[85] Willis, M. C., Palladium-catalyzed coupling of ammonia and hydroxide with aryl halides: The direct synthesis of primary anilines and phenols. *Angewandte Chemie-International Edition* 2007, *46* (19), 3402-3404.

[86] Raphaele, M.; Hohn, D.; Roy, F., Rearrangement of N-methylaniline over H-ZSM-5, H-Theta-1, and H-Y zeolites. *Journal of Catalysis* 1993, *143*, 627-630.

[87] Chang, C. D.; Lang, W. H., Production of aromatic amines using crystalline silicate catalysts. *US Patent 4434299* 1984.

[88] Becker, J.; Niederer, J. P. M.; Keller, M.; Holderich, W. F., Amination of cyclohexanone and cyclohexanol/cyclohexanone in the presence of ammonia and hydrogen using copper or a group VIII metal supported on a carrier as the catalyst. *Applied Catalysis A-General* 2000, *197* (2), 229-238.

[89] Manohar, B.; Ganesh, I.; Reddy, B. M., Aniline synthesis from cyclohexanol and ammonia over mixed oxide catalysts. *Journal of Molecular Catalysis A-Chemical* 1998, *129* (1), L5-L8.

[90] Collet, F.; Lescot, C.; Dauban, P., Catalytic C-H amination: The Stereoselectivity Issue. *Chemical Society Reviews* 2011, *40*, 1926-1936.

[91] Collet, F.; H., D. R.; Dauban, P., Catalytic C-H amination: Recent progress and future directions. *Chemical Communications* 2009, 5061-5074.

[92] Chen, T.; Hu, C. W.; Fu, Z. J.; Tian, A. M., Advance in the synthesis of aromatic amine via direct amination. *Chinese Science Bulletin* 2002, *47* (23), 1937-1939.

[93] Ramirez, T. A.; Baoguo Zhao, B. G.; Shi, Y., Recent advances in transition metal-catalyzed sp3 C-H amination adjacent to double bonds and carbonyl groups. *Chemical Society Reviews* 2012, *41*, 931-942.

[94] Zalatan, D. N.; Boi, J. D., Metal-Catalyzed Oxidations of C-H to C-N Bonds. *Topics in Current Chemistry* 2010, *292*, 347-378.

[95] Bois, J. D., Rhodium-catalyzed C-H amination. An enabling method for chemical synthesis. *Organic Process Research and Development* 2011, *15*, 758-762.

[96] Hili, R.; Yudin, A. K., Making carbon-nitrogen bonds in biological and chemical synthesis. *Nature Chemical Biology* 2006, *6*, 284.

[97] Diversi, P.; Ermini, L.; Ingrosso, G.; Lucherini, A.; Pinzino, C.; Sagramora, L., The reaction of benzene with diethylamine in the presence of an iridium(I) mercury(II) based system - a model route for the direct amination of aromatic-hydrocarbons. *Journal of Organometallic Chemistry* 1995, *494* (1-2), C1-C3.

[98] Stern, M. K.; Hileman, F. D.; Bashkin, J. K., The direct coupling of aniline and nitrobenzene: a new example of nucleophilic aromatic substitution for hydrogen. *Journal of American Chemistry Society* 1992, *114*, 9237-9238.

[99] Huertas, I.; Gallardo, I.; Marquet, J., Direct formation of aromatic C-N bonds. Regioselective amination of m-dinitrobenzene via fluoride promoted nucleophilic aromatic photosubstitution for hydrogen. *Tetrahedron Letters* 2000, *41* (2), 279-281.

[100] Roundhill, D. M., Transition metal and enzyme catalyzed reactions involving reactions with ammonia and amines. *Chemical Reviews* 1992, *92*, 1.

[101] Klinkenberg, J. L.; Hartwig, J. F., Catalytic Organometallic Reactions of Ammonia. *Angewandte Chemie-International Edition* 2011, *50*(1),86-95.

[102] Wibaut, J. P., *Berichte* 1917, *50*, 541-546.

[103] Thomas, C. L., Preparation of aromatic amines. *Canadian Patent 553988* 1958.

[104] Schmerling, L., Preparation of Aromatic Amines. *US Patent 2948755* 1960.

[105] Hara, K., Amination and cyanation of aromatic compounds. *JP Patent 06293715* 1994.

[106] Squire, E. N.; Mills, G., Synthesis of aromatic amines by reaction of aromatic compounds with ammonia. *US Patent 3919155 and 3929889* 1975.

[107] Poojary, D.; Borade, R.; Hagemeyer, A.; Dube, C.; Zhou, Z. P.; Nothels, U.; Armbrust, R.; Rasp, C., Amination of aromatic hydrocarbons and heterocyclic analogs thereof. *PCT WO 0069804* 2000.

[108] Desrosiers, P.; Guan, S. H.; Hagemeyer, A.; Lowe, D. M.; Lugmair, C.; Poojary, D. M.; Turner, H.; Weinberg, H.; Zhou, X. P.; Armbrust, R.; Fengler, G.; Notheis, U., Application of combinatorial catalysis for the direct amination of benzene to aniline. *Catalysis Today* 2003, *81* (3), 319-328.

[109] Hagemeyer, A.; Borade, R.; Desrosiers, P.; Guan, S. H.; Lowe, D. M.; Poojary, D. M.; Turner, H.; Weinberg, H.; Zhou, X. P.; Armbrust, R.; Fengler, G.; Notheis, U., Application of combinatorial catalysis for the direct amination of benzene to aniline. *Applied Catalysis A-General* 2002, *227* (1-2), 43-61.

[110] Hoffmann, N.; Muhler, M., On the mechanism of the oxidative amination of benzene with ammonia to aniline over NiO/ZrO_2 as cataloreactant. *Catalysis Letters* 2005, *103* (1-2), 155-159.

[111] Hoffmann, N.; Loeffler, E.; Breuer, N. A.; Muhler, M., On the nature of the active site for the oxidative amination of benzene to aniline over NiO/ZrO_2 as cataloreactant. *ChemSusChem* 2008, *1* (5), 393-396.

[112] Becker, J.; Hölderich, W. F., Amination of benzene in the presence of ammonia using a Group VIII metal supported on a carrier as catalyst. *Catalysis Letters* 1998, *54* (3), 125-128.

[113] Hölderich, W.; Becker, J., Preparation of amino-substituted aromatic hydrocarbons. *DE Patent 19634110* 1998.

[114] Durante, V. A.; Wijeskera, T. P.; Karmakar, S., Oxidative amination of benzene to anilineUsing molecoular oxygen as the terminal oxidation. *US Patent 5861536* 1999.

[115] Axon, S. A.; Jackson, S. D.; Claes, P. R. R., Process for the Production of Aromatic Amines. *PCT WO 9910311* 1999.

[116] Stitt, H. E.; Jackson, S. D., Amine Production. *PCT WO 0009473* 2000.

[117] Yuzawa, H.; Yoshida, H., Direct aromatic-ring amination by aqueous ammonia with a platinum loaded titanium oxide photocatalyst. *Chemical Communications* 2010, *46* (46), 8854-8856.

[118] Hu, C. W.; Zhu, L. F.; Xia, Y. S. Preparation of catalyst for direct oxidation amination one step synthesizing aniline from benzene. *CN1555921-A*; *CN1283617-C*, 2004.

[119] Chen, T.; Fu, Z. J.; Zhu, L. F.; Hu, C. W.; Tian, A. M., Direct synthesis of aniline by oxido-amination of benzene with H_2O_2 and aqueous ammonia on Ni-Zr-Ce/Al_2O_3 catalyst. *Acta Chimica Sinica* 2003, *61* (11), 1701-1703.

[120] Chen, T.; Zhu, L. F.; Hu, C. W., Study of the Synthesis of Aniline via direct Amination of Benzene with H_2O_2 Oxidation on Zr-Ni/ Al_2O_3. *Journal of Molecular Catalysis (China)* 2005, *19* (4), 275-279.

[121] Xia, Y. S.; Zhu, L. F.; Li, G. Y.; Hu, C. W., Direct oxido-amination of benzene to aniline over nickel-vanadium catalysts. *Acta Physico-Chimica Sinica* 2005, *21* (12), 1337-1342.

[122] Guo, B.; Zhang, Q.; Li, G. Y.; Yao, J. Y.; Hu, C. W., Aromatic C-N bond formation via simultaneous activation of C-H and N-H bonds: direct oxyamination of benzene to aniline. *Green Chemistry* 2012, *14* (7), 1880-1883.

[123] Hu, C. W.; Zhu, L. F.; Xia, Y. S., Direct amination of benzene to aniline by aqueous ammonia and hydrogen peroxide over V-Ni/Al_2O_3 catalyst with catalytic distillation. *Industrial and Engineering Chemistry Research* 2007, *46* (10), 3443-3445.

[124] Mantegazza, M. A.; Leofanti, G.; Petrini, G.; Padovan, M.; Zeccina, A.; Bordiga, S. in: Corberan, V .C.; Bellon, S.V. (Eds.), *New Developments in Selective Oxidation, Elsevier, New York*. 1994; p 51.

[125] Mantegazza, M. A.; Petrini, G.; Fornasari, G.; Rinaldo, A.; Trifiro, F., Ammoximation reaction in the gas and liquid phases with silica based catalysts: Role of titanium. *Catalysis Today* 1996, *32* (1-4), 297-304.

[126] LeBars, J.; Dakka, J.; Sheldon, R. A., Ammoximation of cyclohexanone and hydroxyaromatic ketones over titanium molecular. *Applied Catalysis A-General* 1996, *136* (1), 69-80.

[127] Kuznetsova, N. I.; Kuznetsova, L. I.; Detusheva, L. G.; Likholobov, V. A.; Pez, G. P.; Cheng, H., Amination of benzene and toluene with hydroxylamine in the presence of transition metal redox catalysts. *Journal of Molecular Catalysis A-Chemical* 2000, *161* (1-2), 1-9.

[128] Zhu, L. F.; Guo, B.; Tang, D. Y.; Hu, X. K.; Li, G. Y.; Hu, C. W., Sodium metavanadate catalyzed one-step amination of benzene to aniline with hydroxylamine. *Journal of Catalysis* 2007, *245* (2), 446-455.

[129] Lue, Y. F.; Zhu, L. F.; Liu, Q. Y.; Guo, B.; Hu, X. K.; Hu, C. W., Direct amination of benzene to aniline with several typical vanadium complexes. *Chinese Chemical Letters* 2009, *20* (2), 238-240.

[130] Yu, T. H.; Hu, C. W.; Wang, X. Q., Direct amination of toluene with hydroxylamine in the presence of vanadium-based catalysts. *Chemistry Letters* 2005, *34* (3), 406-407.

[131] Hu, C. W.; Yu, T. H. Direct one-step aminating synthesis of toluidine from toluene. *CN1706808-A*; *CN1706808-B*, 2004.

[132] Hu, C. W.; Yu, T. H. Direct one-step amination process of synthesizing toluidine from toluene. *CN1706807-A*; *CN100551899-C*, 2004.

[133] Hu, C. W.; Zhu, L. F.; Xia, Y. S.; Li, G. Y. One-step method for synthesizing phenylamine from direct oxidation and amination of bezene. *CN1807397-A*; *CN100457712-C*, 2005.

[134] Tang, D. Y.; Zhu, L. F.; Qing, S.; Su, Z. S.; Hu, C. W., A theoretical study on the mechanism of the oxidation of hydroxylamine by VO_2^+. *Journal of Molecular Structure-Theochem* 2007, *805* (1-3), 143-152.

[135] Tang, D. Y.; Zhu, L. F.; Hu, C. W., Comprehensive DFT study of the mechanism of vanadium-catalyzed amination of benzene with hydroxylamine. *Organometallics* 2011, *30* (21), 5675-5686.

[136] Parida, K. M.; Dash, S. S.; Singha, S., Structural properties and catalytic activity of Mn-MCM-41 mesoporous molecular sieves for single-step amination of benzene to aniline. *Applied Catalysis A-General* 2008, *351* (1), 59-67.

[137] Parida, K. M.; Rath, D.; Dash, S. S., Synthesis, characterization and catalytic activity of copper incorporated and immobilized mesoporous MCM-41 in the single step amination of benzene. *Journal of Molecular Catalysis A-Chemical* 2010, *318* (1-2), 85-93.

[138] Singha, S.; Parida, K. M., A reusable Mn(II)-dampy-MCM-41 system for single step amination of benzene to aniline using hydroxylamine. *Catalysis Science and Technology* 2011, *1* (8), 1496-1505.

[139] Gao, L.; Zhang, D.; Wang, Y.; Xue, W.; Zhao, X., Direct amination of toluene to toluidine with hydroxylamine over $CuO-V_2O_5/Al_2O_3$ catalysts. *Reaction Kinetics Mechanisms and Catalysis* 2011, *102* (2), 377-391.

[140] Lisitsyn, Y. A.; Kargin, Y. M., Electrochemical amination of benzene in aqueous solutions of sulfuric acid and an organic solvent. *Russian Journal of Electrochemistry* 2003, *39* (12), 1279-1284.

[141] Lisitsyn, Y. A.; Kargin, Y. M., Electrochemical amination in the system $Ti(IV)-NH_2OH-C_6H_6$: The efficiency of the process in a sulfuric acid solution containing an organic solvent. *Russian Journal of Electrochemistry* 2004, *40* (8), 864-867.

In: Aniline
Editors: K. Hernandez, M. Holloway

ISBN: 978-1-62417-417-9

Chapter 4

SYNTHESIS AND STRUCTURES OF GOSSYPOL MONO-ALDEHYDE DERIVATIVES WITH O-SUBSTITUTED ANILINES

K. Z. Tilyabaev*, F. G. Kamaev, A. M. Yuldashev and B. T. Ibragimov
Institute of Bioorganic Chemistry,
Uzbek Academy of Sciences,
Tashkent, Uzbekistan

ABSTRACT

Gossypol mono-aldehyde derivatives with *o*-methylaniline, and *o*-hydroxyaniline have been synthesized and their structures have been studied by IR and ^{1}H NMR spectroscopic methods.

Keywords: Gossypol, o-methylaniline, o-hydroxyaniline, synthesis, structure, enamine tautomer

* Institute of Bioorganic Chemistry, Uzbek Academy of Sciences, Mirzo Ulugbek str., 83, Tashkent, 100125, Uzbekistan, E-mail:tilyabaevzaid@mail.ru.

Introduction

Aniline and its analogs are primary aromatic amines that form Schiff bases via reaction with aldehydes. They are known to exist as imine and enamine tautomers [1]. The reaction of aniline with gossypol (Figure, A), a yellow aldehydonaphtholic pigment of cottonseed, gives well-crystallized product called dianilinegossypol [2], which exists in enamine form in a solution and in a solid state [3]. Reactions of gossypol with aniline and antranilic acid are used to determine gossypol content as well as to isolate it from cottonseed processing products [4]. The most of condensation products of gossypol with aromatic amines are bis-derivatives. However, there are some data on gossypol mono-derivatives with aniline [5], *o*-phenylenediamine [6] and *o*-nitroaniline [7]. Monoanilinogossypol has been found to exhibit antiradical [8], antifungal [9], anticancer [10] activities accompanied with low acute toxicity [5]. This makes unsymmetrical derivatization of gossypol molecule very promising in terms of potential drug development and allows obtaining by this principle numerous gossypol derivatives with aromatic amines that may demonstrate similar or even better effect. In this chapter we report on synthesis and structure elucidation of gossypol half-enamines with substituted anilines such as *o*-methylaniline, and *o*-hydroxyaniline.

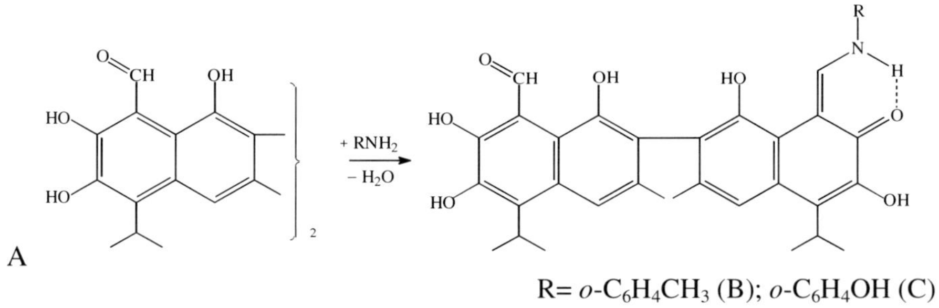

Figure. The general scheme of the preparation of gossypol half-enamines with *o*-methylaniline (B), and *o*-hydroxyaniline (C).

Results and Discussion

The general scheme of the preparation of gossypol half-enamines with *o*-methylaniline (B), and *o*-hydroxyaniline (C) is given in Figure. In order to synthesize gossypol mono-aldehyde derivatives B and C a homogeneous reaction was used. For minimizing the possibility of both gossypol aldehyde groups to react rapidly with aminocompounds we applied high dilution technique, so amine was added to diluted gossypol solution, which was cooled as well as the solution of amine. A spectral criterion of the completeness of the mono-derivative formation was the presence in the ^{1}H NMR spectra of resonances of aldehyde protons at 11.2–11.3 ppm and =C<u>H</u>-NH- protons at 10.2–10.5 ppm at a 1: 1 ratio.

Structural investigations of newly synthesized gossypol mono-aldehyde derivatives by ^{1}H NMR spectroscopy method have demonstrated that they (to put more precisely, their amino-substituted part) exist predominantly as enamine tautomer. The enamine structure of the synthesized compounds was suggested based on the appearing doublet at 10.2–10.5 ppm resulting from the spin–spin interaction of methyledene and amine protons typical for enamine forms [11].

EXPERIMENTAL PART

Gossypol (A) was prepared at an experimental plant of the Institute of Bioorganic Chemistry, Uzbek Academy of Sciences, from side products of cottonseed oil production. Chemical reagents and solvents were obtained from Aldrich. The melting points were measured in a glass capillary. TLC was carried out on Silufol UV-254 plates with the development in an iodine chamber. The IR spectra were recorded on an IR-10 spectrophotometer in the region of 3700–750 cm^{-1} in KBr tablets. The ^{1}H NMR spectra ($CDCl_3$ and acetone-d_6) were registered on NMR-UNITY 400+ spectrometer with a working frequency of 400 MHz and tetramethylsilane as an internal standard.

8-Formyl-1,1,6,6',7-pentahydroxyl-3,3'-dimethyl-8'-ilidenamino-[2''-toluyl]-7'-oxo-5,5'-diisopropyl-2,2'-binaphthalene (B). Solutions of compound (A) (104 mg, 0.2 mmol) in chloroform (3 ml) and o-methylaniline (22 μl, 0.2 mmol) in chloroform (1 ml) were kept for 1 day at –15°C and then mixed and kept for several days at the same temperature. The reaction was monitored by TLC (elution with chloroform, R_f =0.32 for compound (B). The reaction mixture was concentrated and applied on silica gel (100/160 μm, elution with chloroform) to give 60% (B); mp 275-277°C. Found, %: C 72.91; H 5.92; N 2.28. $C_{37}H_{37}N_1O_7$. Calc., %: C 73.15; H 6.10; N 2.31. IR (KBr), ν, cm^{-1}: 3400-3500 (O-H, N-H), 1540 (C=O). ^{1}H NMR: (400 MHz, $CDCl_3$): 15.38 (1 H, s, C7–*OH*), 15.20 (1H, b.d., =CH–N*H*), 11.25 (1 H, s, –*C*HO), 10.25 (1 H, d, =C*H*–NH), 7.80 (1H, s, C4-*H*), 7.70* (1 H, s, C4-*H'*), 7.15–7.40* (4H, m, substituent's Ph), 3.60–4.05 (2 H, m, C5–C*H*(CH_3)$_2$), 2.53 (3H, s, substituent's Ph-CH_3), 2.20 (3 H, s, C3–CH_3), 2.15* (3H, s, C3–CH_3'), 1.60* (6H, d, -CH(CH_3)$_2$). The C1–*OH* and –C1'–*OH*, as well as the C6–*OH* and –C6'–*OH* protons could not be identified separately due to proton exchange. Herein and below, the resonances of the substituted fragments are marked with an asterisk (*). 8-Formyl-1,1,6,6',7-pentahydroxyl-3,3'-dimethyl-8'-iliden amino-[2''-hydroxyphenyl]-7'-oxo-5,5'-diisopropyl-2,2'-binaphthalene (C). Solutions of compound (A) (104 mg, 0.2 mmol) in acetone (3 ml) and o-hydroxyaniline (22 mg, 0.2 mmol) in acetone (1 ml) were kept for 1 day at –15°C and then mixed and kept for several days at the same temperature. The reaction was monitored by TLC (elution with benzene: ethyl acetate 9:1(v/v), R_f =0.51 for compound (C). The reaction mixture was concentrated and applied on silica gel (100/160 μm, elution with benzene ethyl acetate 9:1 (v/v)) to give 61% (C); mp 273-276°C. Found, %: C 70.81; H 5.62; N 2.28. $C_{36}H_{35}N_1O_8$. Calc., %: C 70.94; H 5.75; N 2.30. IR (KBr), ν, cm^{-1}: 3500-3600 (O-H, N-H), 1590 (C=O). ^{1}H NMR: (400 MHz, acetone-d_6): 11.33 (1 H, s, –C*HO*), 10.50 (1 H, d, =C*H*–NH), 7.82 (1H, s, C4-*H*), 7.60* (1 H, s, C4-*H*'), 6.65–7.50* (4H, m, substituent's Ph), 3.60–4.10 (2 H, m, C5–C*H*(CH_3)$_2$), 2.18 (3 H, s, C3–CH_3), 2.12* (3H, s, C3–CH_3'), 1.55 (6H, d, -CH(CH_3)$_2$), 1.48* (6H, d, -CH(CH_3)$_2$). Signals of NH- and OH-protons could not be indentified separately due to the presence of water in the solvent.

CONCLUSION

Thus, gossypol mono-aldehyde derivatives with *o*-methylaniline and *o*-hydroxyaniline have been synthesized. ^{1}H NMR spectral results allow suggesting enamine structure for amino-substituted part of the molecules synthesized.

REFERENCES

[1] Simkin, B. Ya, Bran', V.A., Minkin, V.I. *Russian J. Org. Chem.* 1977, 13, 1710-1722.

[2] Matlin, S.A., Zhou, R. J. High Resolut. *Chromatogr., Chromatogr. Commun.* 1984, 7, 629–631.

[3] Nazarov, G.B., Ibragimov, B.T., Aripov, T.F. *Chem. Nat.Compd.* 1988, 24, 560–564.

[4] Markman, A.L., Rzhekhin, V. P. Gossypol and Its Derivatives. Moscow: Pishchevaya Promyshlennost, 1965, 246 pp. [Engl. translation: *Israel Program for Scientific Translations*, Jerusalem 1969].

[5] Tilyabaev, K.Z., Kamaev, F.G., Vypova, N.L., Yuldashev, A.M., Ibragimov, B.T., Talipov S.A. *Russian J. Bioorg. Chem.* 2010, 36, 390-395.

[6] Karrer, P., Tobler, E. *Helv. Chim. Acta.* 1932, 15, 1204–1212.

[7] Dechary, J.M., Brown, L.E. *J. Am. Oil Chem. Soc.* 1956, 33, 76–78.

[8] Gayibov U.G., Tilyabaev K.Z., Salakhutdinov B.A., Yuldashev A.M., Talipov S.A., Ibragimov B.T. *Uzbek biological journal.* 2009, 3, 3 – 6.

[9] Tilyabaev, K.Z., Arzanova, I.A., Yuldashev, A.M., Ibragimov, B.T., and Talipov, S.A. *Uzbek biological journal.* 2009, 4, 7 – 10.

[10] Tilyabaev, K.Z., Yuldashev, A.M., Ibragimov, B.T., Talipov, S.A. in: Proceedings of Scientific conference "Biologically active compounds: fundamental aspects of obtaining and application", *Novy Svet.* 2009, 431-432.

[11] 11 Baram, N.I., Ismailov, A.I., Biktemirov, L., Paizieva, R.Z., and Ziyaev, Kh.L. in *Problemy i perspektivy razvitiya khimii prirodnykh i fiziologicheski aktivnykh veshchestv* (Problems and Prospects of Development of Chemistry of Natural and Physiologically Active Compounds), Tashkent: Fan, 1988, pp. 78–99.

In: Aniline
Editors: K. Hernandez, M. Holloway
ISBN: 978-1-62417-417-9

Chapter 5

OXIDATIVE POLYMERIZATION OF DIPHENYLAMINE-2-CARBONIC ACID –AN AROMATIC DERIVATIVE OF ANILINE

S. Zh. Ozkan**, I. S. Eremeev and G. P. Karpacheva
Russian Academy of Sciences A.V. Topchiev Institute of Petrochemical Synthesis, Moscow, Russia

ABSTRACT

Polymers of diphenylamine-2-carbonic acid (DPhACA) were obtained for the first time by means of chemical oxidative polymerization in aqueous solution of ammonium hydroxide, a new polyacid, containing carboxyl groups which form intermolecular hydrogen bonds with aminogroups along the whole polymer chain, was obtained. The effect of polymerization conditions on the chemical structure and molecular-mass characteristics of polydiphenylamine-2-carbonic acid (PDPhACA) was studied.

According to IR spectroscopy data, while polymerization of DPhACA goes in ammonium hydroxide solution, growth of polymeric chain occurs due to C–C joining into 2- and 4- positions of aromatic rings

* E-mail: ozkan@ips.ac.ru.

relative to nitrogen. DPhACA polymers are amorphous and thermally stable.

Keywords: Oxidative polymerization, polymers of diphenylamine-2-carbonic acid, UV and IR spectroscopy

Interest for the polymers based on aniline and its derivatives increases every year, it can be explained by simplicity of the synthesis under conditions of oxidative polymerization, high stability in operation and a complex of excellent physicochemical properties [1-5]. However, expansion of the range of electroactive polymers is being retarded by the limited solubility of new monomers [6-13] in standard conditions of oxidative polymerization (in aqueous solutions of acids).

In the current work polymers of diphenylamine-2-carbonic acid (DPhACA) – an aromatic derivative of aniline were obtained in the aqueous solution of ammonium hydroxide for the first time.

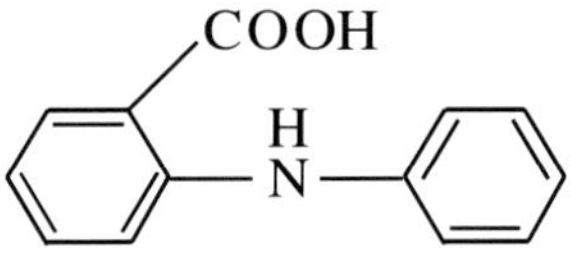

Oxidative polymerization of DPhACA is impossible to conduct in standard conditions, because it does not dissolve in aqueous acid medium. Since the monomer is soluble in aqueous alkaline solutions, oxidative polymerization of DPhACA was carried out in the solution of ammonium hydroxide for the first time. Analysis of the effect of reaction conditions (concentration of monomer, oxidizer, ammonia, temperature and time of the synthesis) on the yield and molecular-mass characteristics of PDPhACA has shown the following: yield of the polymer grows with the increase of monomer concentration.

To obtain products with the highest molecular weight, the polymerization of DPhACA in the solution of NH_4OH is preferably to be carried out at sufficiently high concentrations of oxidizer (0.2–1.0 mol/l) and ammonia (1 mol/l).

It was found that the polymerization of DPhACA in the solution of NH_4OH can be carried out for a long time without destroying the oxidation products. The best results are achieved when the polymerization of DPhACA is held for 3–24 hours. The temperature of the reaction mixture must be maintained in the range of 0 – +15°C. In this case the molecular weight of the polymer reaches $M_w = 1.9 \times 10^4$, the degree of polymerization is 90, polydispersity index is 2.0.

DPhACA polymer is a black powder, fully soluble in aqueous solutions of NH_4OH and NaOH, N-methylpyrrolidone, DMF, DMSO, partially in THF, dioxane, acetone. As one can see, the presence in the structure of PDPhACA of active carboxyl groups makes it soluble in aqueous alkaline solutions in the same way as it is in the case with the introduction of carboxyl groups in the aromatic ring of polyaniline. According to X-ray analysis, PDPhACA is an amorphous polymer.

Structure of DPhACA polymers was studied by IR and UV spectroscopy. Absorption bands at 825 and 784 cm^{-1} in the IR spectra of polymers (Figure 1) are due to nonplanar deformation vibrations of bonds $\delta_{C\text{–}H}$ of 1,2,4- and 1,2-substituted benzene ring. It shows that during the polymerization of DPhACA in the ammonium hydroxide solution the growth of polymer chain occurs due to C–C joining into 2- and 4-positions of the phenyl rings relative to nitrogen.

In this case carboxyl groups along the whole polymer chain form intramolecular hydrogen bonds with aminogroups, it was confirmed by the appearance of band in the region of 3288 cm^{-1} in the IR spectra and band λ_{max} = 550 nm (n-π * - transition) in the electronic absorption spectra of polymers (Figure 2).

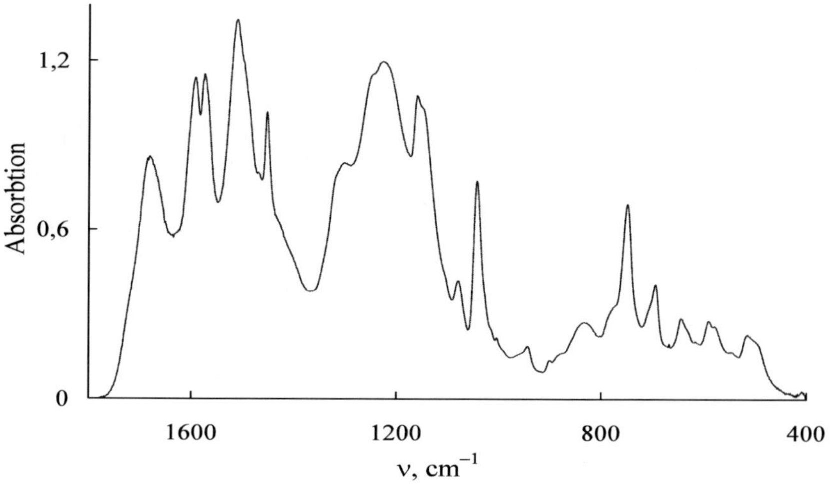

Figure 1. IR spectrum of PDPhACA.

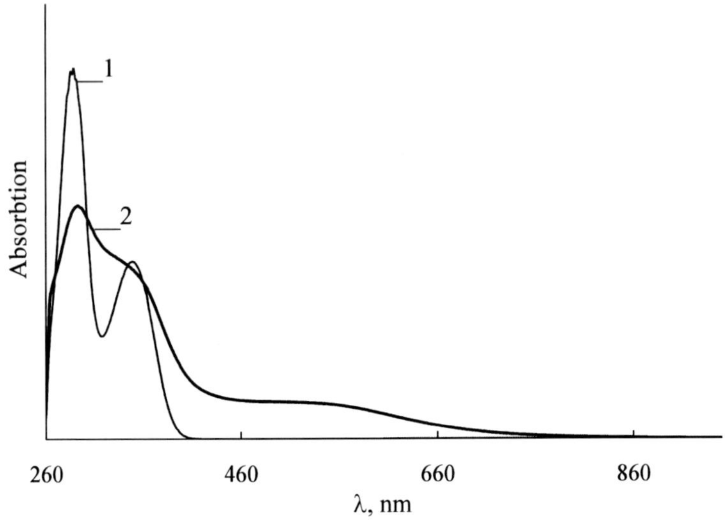

Figure 2. Electronic absorption spectra of monomer (1) and PDPhACA (2) in DMF.

Analysis of the results of spectral studies provides us with the chemical structure of PDPhACA, obtained by polymerization in the aqueous solution of ammonium hydroxide:

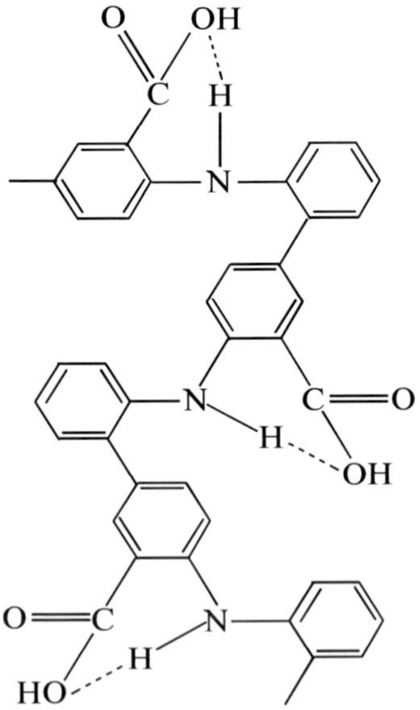

Figure 3. (Continued).

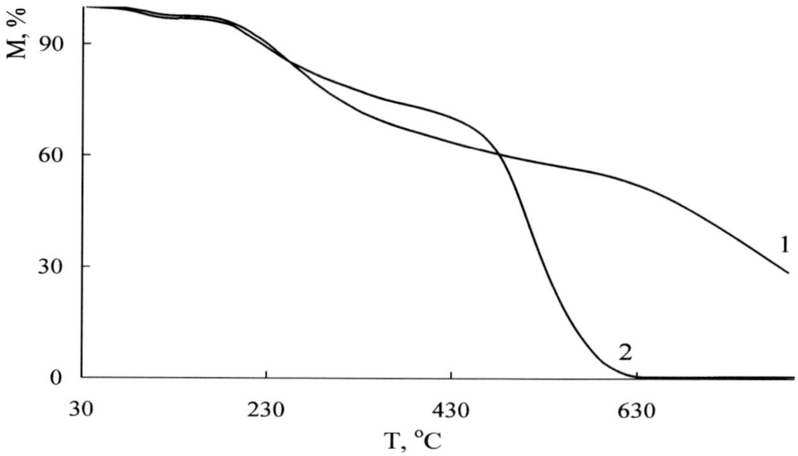

Figure 3. The loss in weight of PDPhACA upon heating to 800°C at the rate 10 deg/min in the flow of nitrogen (1) and in the air (2).

The presence in the structure of PDPhACA of two joined phenyl rings results in high thermal stability of the polymer. Loss of weight at 168°C is due to the removal of COOH groups (Figure 3).

Such behavior is also typical for PDPhACA obtained in sulfuric acid solution [12]. There is an exothermic peak associated with decomposition on the DSC thermogram of PDPhACA in this temperature range. When heated once again, this peak is absent.

Removal of COOH groups was confirmed by high-temperature IR spectroscopy. Comparative analysis of the IR spectra of the original polymer and the polymer heated to 200°C in the air showed that with the increase of temperature there is gradual reduction of the intensity of the bands at 1677 and 1239 cm^{-1}, which characterize the COOH groups. It was found that after complete removal of the COOH groups the polymer structure is equal to the structure of polydiphenylamine.

The main processes of thermooxidative destruction of PDPhACA start at 470°C. One can observe the 5% weight loss for PDPhACA at 197°C, and 50% weight loss – at 520°C.

It is known that 50% weight loss for polyaniline is observed at 450°C. In the inert atmosphere PDPhACA loses half of the original weight at 660°C. At 800°C the remainder is 28%.

REFERENCES

[1] Karyakin A.A., Vagin M.Yu., Ozkan S.Zh., Karpachova G.P. // *J. Phys. Chem. B.* 2004. V. 108. № 31. P. 11591.

[2] Vagin M.Yu., Trashin S.A., Ozkan S.Zh., Karpachova G.P., Karyakin A.A. // *J. Electroanal. Chem.* 2005. V. 584. № 2. P. 110-116.

[3] Trashin S.A., Vagin M.Yu, Karpachova G.P., Ozkan S.Zh., Karyakin A.A. // *J. Nano and Microsystem Technique*. 2008. № 8. P. 49.

[4] Ozkan S.Zh., Kozlov V.V., Karpacheva G.P. // *Journal of the Balkan Tribologcal Association.* 2010. Book 3. V. 16. № 3. P. 393.

[5] Ozkan S.Zh., Dzidziguri E.L., Karpacheva G.P., Bondarenko G.N. // *Nanotechnologies in Russia.* 2011. V. 6. № 11-12. P. 750.

[6] Karpacheva G.P., Orlov A.V., Kiseleva S.G., Ozkan S.Zh., Yurchenko O.Yu., Bondarenko G.N. // *Rus. J. Electrochem.* 2004. V. 40. № 3. P. 305.

[7] Orlov A.V., Ozkan S.Zh., Bondarenko G.N., Karpacheva G.P. // *Polym. Sci. B.* 2006. V. 48. № 1-2. P. 5.

[8] Orlov A.V., Ozkan S.Zh., Karpacheva G.P. // *Polym. Sci. B.* 2006. V. 48. № 1-2. P. 11.

[9] Ozkan S.Zh., Karpacheva G.P., Orlov A.V., Dzyubina M.A. // *Polym. Sci. B.* 2007. V. 49. № 1-2. P. 36.

[10] Ozkan S.Zh., Bondarenko G.N., Orlov A.V., Karpacheva G.P. // *Polym. Sci. B.* 2009. V. 51. № 5-6. P. 149.

[11] Korolev Yu.M., Ozkan S.Zh. // *Doklady Physical Chemistry*. 2009. V. 429. Part 1. P. 223.

[12] Ozkan S.Zh., Bondarenko G.N., Karpacheva G.P. // *Polym. Sci. B.* 2010. V. 52. № 5. P. 263.

[13] Ozkan S.Zh., Karpacheva G.P., Bondarenko G.N. // *Rus. Chem. Bull.* 2011. № 8.

INDEX

A

B

C

D

E

F

G

H

I

J

K

L

M

N

O

P

R

S

T

U

V

W

X

Y

Z